THE STEPHEN BECHTEL FUND

IMPRINT IN ECOLOGY AND THE ENVIRONMENT

The Stephen Bechtel Fund has

established this imprint to promote

understanding and conservation of

our natural environment.

The publisher gratefully acknowledges the generous contribution to this book provided by the Stephen Bechtel Fund.

DIRTY
WATER

DIRTY
WATER

**One Man's Fight
to Clean Up One of the World's
Most Polluted Bays**

Bill Sharpsteen

UNIVERSITY OF CALIFORNIA PRESS

Berkeley Los Angeles London

University of California Press, one of the most distinguished university presses in the United States, enriches lives around the world by advancing scholarship in the humanities, social sciences, and natural sciences. Its activities are supported by the UC Press Foundation and by philanthropic contributions from individuals and institutions. For more information, visit www.ucpress.edu.

University of California Press
Berkeley and Los Angeles, California

University of California Press, Ltd.
London, England

Unless otherwise noted, all photographs are by the author.

Library of Congress Cataloging-in-Publication Data
Sharpsteen, Bill.
    Dirty water : one man's fight to clean up one of the world's most polluted bays / Bill Sharpsteen.
        p.   cm.
    Includes bibliographical references and index.
    ISBN 978-0-520-25660-6 (cloth : alk. paper)
        1. Sewage disposal in the ocean—Environmental aspects—California—Santa Monica Bay.   2. Bennett, Howard, 1929–   3. Hyperion Water Treatment Plant (Los Angeles, Calif.)—History.   4. Marine pollution—California—Santa Monica Bay—Prevention.   5. Santa Monica Bay Region (Calif.)—Environmental conditions. 6. Environmental protection—Government policy—California—Santa Monica Bay Region.   7. Environmentalists—United States—Biography.   8. Political activists—United States—Biography.   9. High school teachers—United States—Biography.   I. Title.
    TD763.S557 2010
    363.739'409794—dc22   2009003365

Manufactured in the United States of America

19 18 17 16 15 14 13 12 11 10
10 9 8 7 6 5 4 3 2 1

This book is printed on Cascades Enviro 100, a 100% post consumer waste, recycled, de-inked fiber. FSC recycled certified and processed chlorine free. It is acid free, Ecologo certified, and manufactured by BioGas energy.

For Gloria.
You prodded, you encouraged, you offered advice. . . .
You were so patient.

Contents

Illustrations

Acknowledgments

In some ways, this book is an oral history, a collage of people's memories assembled into a story. Naturally, the book wouldn't have been possible had those involved in the story refused my interview requests. In fact, several agreed to subsequent interrogations, patiently answering my follow-up questions so I could be as accurate as possible in retelling their stories. And so: all those who took the time to sit down with me, I truly appreciate your contributions.

In particular, Howard Bennett was one of my biggest sources for information. He freely opened his files to me, allowing me to use whatever I wanted. Given that he apparently saved every scrap of paper related to the "campaign," as he called it, I got an inside look into the story that would have been impossible with only interviews and newspaper articles. Bennett also managed to collect nearly every television report about him or his campaign during that time, and he gave me copies, which rounded out my research in ways that would have otherwise been impossible.

David Brown was another wonderful source of material. We spent hours talking about his story, and he, too, had saved just about everything he collected during his days at SCCWRP. In a fit of frustration, he once tossed a lot of it in the garbage, but his far more prescient wife, Anne, dug through the trash can and recovered the records that would much later be so useful to me. Brown shared with me anything I needed to get deep into the story of SCCWRP and Willard Bascom, providing me with dozens and dozens of newspaper articles and television stories on the subject and saving me a huge amount of research time.

Other major contributors of information were Dorothy Green, Mark Gold, Don May, Ed Tarvyd, Felicia Marcus, John Dorsey, Leif Bennett, Mas Dojiri, Maureen Kindel, Tom Hayden, Robert Ghirelli, and Moe Stavnezer, among the many people I talked with.

One difficulty in telling a story that occurred twenty-three years before I started researching it is being able to describe a scene with any kind of depth. California's Los Angeles Regional Water Quality Control Board helped tremendously here by digging through who knows how many dusty boxes before finding the original reel-to-reel tapes from their hearings regarding the 301(h) waiver. I was able to hear Rim Fay's voice recorded on March 25, 1985 (apparently the meeting transcripts were nowhere to be found). In addition, I almost felt as though I had a seat at the crowded May 13, 1985, meeting and could listen to each speaker address the waiver. My listening to these tapes was hampered by a few technical issues, such as the obsolescence of the equipment and the tape speed, and I would have heard nothing more than a high-pitched squeal without the help of audio wizard Rodney Pearson.

Thanks also to Julie Popkin, who helped secure my publishing contract with University of California Press, and of course, to the people at UC Press who helped me put this book together, in particular my editor, Jenny Wapner, her ever-cheerful assistant Lisa Tauber, copyeditor Bonita Hurd, and project manager Laura Harger.

Author's Note

It's a truism in journalism that most people want to tell their stories. All you have to do is ask and they will gladly pour out their memories. Unfortunately, while the events depicted in this book are hardly ancient history, they're old enough for the people involved to be a little shaky on the facts and, in particular, the chronology. It's gotten to the point where certain official accounts of these events written years later are mostly truth spiced with a little myth. I found this to be so partly because certain versions of this story have been verbally passed around without scrutiny and told as if they were solid, factual history, which they were not. Even the people who were present at the events know only half of what really happened.

This makes telling the story based on interviews with the participants a tad difficult. Having said that, let me point out that I've done everything I can to verify with second and sometimes third sources the stories I relate in this book. To determine the chronology, I pieced together what happened when by means of newspaper articles, dated press releases, and government records. But because so much of the story is a personal one, and the official records rarely relate the emotional aspects, I've had to trust that certain accounts are true. And when people said they couldn't remember something specific but then guessed at what happened, I left their recollections out altogether.

So is this an excuse for sloppy research? Nope. I just want the reader to understand that 95 percent of what I relate in this book is verified and factual, and a smidgen of it is perhaps someone's version filtered through years of faded memories. And in cases of the latter, I point out the difference between fact and slightly obscured truth. That is, after all, a description of history, which depends on witness accounts that should be suspect even if they are interesting to read.

Lastly, if you have any comments or questions, or would like to arrange a reading or interview, feel free to write me at dirtywater1@yahoo.com.

Surfer Scientist

Dr. John Dorsey liked to call it black mayonnaise. That pretty much described the thick mat of sewage sludge that lay on the seafloor some 320 feet below him as he hauled up a sediment sample from the area called Site 8A. The *Marine Surveyor,* the twenty-year-old boat he had taken to this point seven miles off-shore, barely rocked on the early summer seas, and it seemed as though the Pacific Ocean that surrounded him was pure, clean, and untouched.

It wasn't really. Some people called that sludge below him a dead zone, an underwater desert, which wasn't actually lifeless but was devoid of most of the diverse marine life that once had lived there. Only a few species remained. Near here was the out-fall of a seven-mile pipe from Los Angeles' Hyperion Sewage Treatment Plant that disgorged the city's sewage solids, called sludge, after they had been separated from the rest of the waste-water in large tanks. In 1984, an average 4.8 million gallons per day were pumped through this pipe—which had first started operating thirty-three years before—a total of 49,414 tons for the year. The sludge was so thick and heavy it just accumulated on the seafloor, smothering the life that was once there.

To be clear, this wasn't the only source of sewage flowing into the bay. A five-mile pipe discharged the wastewater that remained after the sludge was removed. And a one-mile pipe relieved Hyperion of untreated sewage during plant malfunctions or when the volume reached such immense levels that the facility couldn't handle it all.

This was May 1985. Dorsey, a tall young man who was as happy surfing as he was researching the ocean, had been hired by Hyperion in 1983, after graduating from the University of Melbourne the year before with a PhD in marine biology and pollution ecology (he went to the school because of both a scholarship and the nearby surfing). The advanced degree made him unique among his peers at Hyperion, and so while his primary job may have involved collecting and analyzing data, he was also slated on occasion to testify before various local government entities, such as the city council, and at an upcoming hearing before the state's Regional Water Quality Control Board. There, as he understood it, some agitated protesters were about to declare that the entire Santa Monica Bay, where that seven-mile outfall sat, was polluted, a giant cesspool, because of Hyperion. He figured he could bring a little balance to the clamor by showing that many parts of the bay were still in decent shape. A scientist just presenting the facts.

Then again, there was that dead zone of sludge below him. He had just finished cowriting Hyperion's 1984 report on Santa Monica Bay's environmental health, which was the first time anyone at the plant had bothered to take the data about the sludge and other nearby marine environments and sift it through a scientist's well-educated brain. Since 1971, Dorsey's employer, the Los Angeles Bureau of Sanitation, which ran Hyperion, had collected raw numbers on sea life and water quality, but it had all sat in files, unexplained. And now, in scientists' plain, dry prose, Dorsey and his coauthors had written, "Embedded within the changed area of sea-bottom was a degraded area around the terminus of the 7-mile sludge outfall. Few species occurred here, resulting in very low diversity, abundance, and biomass." In other words, a dead zone. The biomass, or the total number of living things, had been buried by tons of sludge. He also found elevated levels of cadmium, chromium, and copper. "Sediments at the terminus of the 7-mile outfall were the most severely impacted with concentrations of all four metals ranging from 15 to 65 times background levels," the report said. Indeed, Dorsey told whoever would listen that the discharge of sludge had to stop (by the way, Los Angeles was required by law to find another, less environmentally harmful place for the sludge, but was continuing the discharge because many in the city government believed the black mayonnaise actually benefited marine life).

Nearly as bad was the five-mile outfall pipe, where most of the wastewater landed, about 404 million gallons a day, after the solids had been removed. "Around 25 to 35 square kilometers of sea-bottom around these outfalls (or approximately 8% of the Bay's area) has a macrofaunal assemblage whose structure has been changed by wastewater discharges," Dorsey wrote, referring to the visible marine life. "You could tell it was a real polluted area," he says now, "because it didn't have a lot of species and the ones that were there were in pretty large numbers. So it's like a weed field. It had become analogous of a vacant lot."

It's not that this was news to other marine biologists in the area, but until lately, the city as a whole had had no idea that what was being flushed down its toilets and what was running down its sink drains (about 100 gallons per person per day) had turned nearly a tenth of the bay into the underwater equivalent of a landfill. Instead, Santa Monica Bay was simply the place where 45 million people came each year to swim, surf, boat, or fish. It was part of Los Angeles' identity, a major slice of its personality. Going west to the ocean meant more than just recreation: it revitalized one's enthusiasm for life after all the traffic, crowds, and pollution had sucked one nearly dry.

Santa Monica Bay wasn't really a bay in the classic sense of a mostly enclosed body of water. It was more like a bite-shaped dent in the coast that covered 565 square miles and had more than 50 miles of beaches and assorted piers. It was part marine wonder, part carnival. And even as offensively close as it was to acres of parking lots, restaurants, and highways, the bay's size overpowered all that, and it felt as if it were just as natural as it had been a hundred years ago. The water appeared so pure that people rarely worried about jumping in to cool the skin or wash the soul. If nothing else, all one had to do was turn one's back to the east and face the ocean. One could easily fool oneself into believing nothing else existed.

Out there in the *Marine Surveyor*, Dorsey could see only water, and for a few hours the city and Hyperion indeed disappeared. The people who were shouting about how polluted the bay was couldn't reach him. Perhaps bravely, he was about to tell them at the public hearing that things weren't as bad as they seemed. After all, as his report remarked, "aside from these changed areas around the outfalls, much of the remaining areas in the Bay have richly diverse macrofaunal assemblages characteristic of shelf habitats in southern California."

He wasn't trying to distort or ignore the facts, but in his mind the situation wasn't nearly as apocalyptic as environmentalists were saying. He even surfed on his lunch breaks near Hyperion's third outfall, which extended just one mile offshore and, although designated for emergencies only, discharged a daily average of 2.9 millions gallons of chlorinated effluent.

In the end, Dorsey was a minor player in the story that follows, and that didn't hurt his feelings at all. He preferred surfing to public testimony, and besides, this was one of those issues where expressing the scientific shades of gray attracted few listeners. Instead, the extremes came through loudest. Some said the bay's pollution was the worst in the world. A few argued that, taken as a whole, the sewage from Hyperion wasn't really destroying the beloved bay. It was this view, which ignored the dead zone, with its black sludge, that prevailed until a lone swimmer, shivering in the water on an early spring morning, discovered what a lie that was.

CHAPTER I

The Swimmer

Howard Bennett swam every day in nothing but an old, stretched-out black Speedo, dunking his slender, mostly naked body into the frigid, fifty-six-degree Pacific Ocean at six o'clock in the morning. Given his habit of self-deprecation, he might have admitted how stupid this was, especially in winter, but Bennett *had* to swim each morning. This was more than mere routine; it approached a biologic need. He symbolically, almost ritually, washed off the stress he had accumulated as a high school teacher the day before by stroking away in the dark morning water. He *needed* to be enveloped in that chilling water for the twenty-minute-or-so swim just beyond the surf line. He couldn't imagine a day without it. It renewed him. It was better than a hammock. It was better than sleeping in a few more minutes.

Except for days when the weather was so bad the thrashing waves would have killed him before the cold could, he slipped out of bed when the alarm's buzzer woke him and pulled out the Speedo he kept in the nightstand. He used to run, but in his thirties his toe joints had developed an inflammation so bad it couldn't be fixed by rest or surgery, and so a Scottish surgeon had recommended swimming. "I don't know how to swim," Bennett had told him, as if a new routine scared him. "Then damn it, learn," the doctor said.

Bennett joyfully tells this story now as a series of punch lines timed with such ease that you're both captivated by his delivery and a little suspicious of the tale's veracity. In other words, he's a consummate storyteller. Just the same, he leaves out just how important this moment really was. The doctor's suggestion, possibly thrown off in half-jest, affected not only Bennett's life but also, one could say—with a touch of the hyperbole he often employs himself—an entire city. It could even be argued that Bennett banging away out there every morning in his skimpy swim trunks was the catalyst for changing people's careers, in both good ways and bad, and that it cost Los Angeles billions of dollars. Not bad for a guy who was just trying to relax before going to work.

As it turned out, Bennett fell in love with the ocean and the skilled simplicity of the crawl stroke he soon learned. In 1961, he and his wife, Bente, bought a boxy two-story house on Playa del Rey beach, south of Los Angeles, where they rented out the first floor and kept the second-floor view of the beach for themselves. Having Santa Monica Bay as essentially their backyard was a grand thing. It was so big, so powerful, it belittled anyone who thought of it as their own; and yet, after a few years of living on the beach and learning both the pleasures and dangers of swimming in the impersonal ocean, Bennett couldn't help feeling as though he had an intimate relationship with the bay. As big as it was, he wanted to protect it.

On the morning of March 28, 1985, Bennett pulled his swim-
suit over his slender legs, tied its floppy waistband tight around
his flat stomach, and walked down the long hall from his bedroom to the guest bathroom, where his swim goggles hung on
the end of a towel rack. Bente, still in bed, listened to him for
a moment as he walked out the front door and, with bare feet,
quietly descended the outside stairs to their small backyard garden, where he exited to the beach.

The sun had risen about fifteen minutes earlier in a partly
cloudy sky, but it hadn't yet topped the low hills to the east, so
the shadowless dawn still hung over the hundred yards of beach
between Bennett's house and the whooshing surf. For Bennett,
whose eyesight could best be described as a notch or two better
than blindness, the scene came through as a muted blur of soft
shapes. But then, the great thing about an empty beach was: you
couldn't trip over anything, and even if you did, it wouldn't
hurt. So he'd left his glasses back in the house.

After so many years of swimming, there was little fat on Bennett's nearly six-foot frame. His chest looked strong, the pecto-
ral muscles well defined. In a business suit, standing before his
classes, he looked older, a curmudgeon with thinning hair, but
this morning, he seemed half his age, a tall, wiry jock still feeling his youth.

It was the kind of early-spring morning when the damp,
forty-degree air was colder than the wet sand or, for that matter, the water, so he jogged out to get into the warmer ocean as
soon as he could. As the waves got louder, he saw an old man at
the edge of the surf sitting on a three-legged stool sunk into the
sand, with a lit Sterno can and fishing pole.

"Don't go in water! No swim! Very bad for you! Poison!" the man shouted as Bennett approached the shoreline.

Without his glasses, Bennett thought the guy looked Japanese. Well, Asian, at least. Bennett had seen him fishing most mornings and assumed he lived in the hills above the beach, where the houses were bigger and more elegant than the plain beach homes stacked side by side.

Given that this was the first time either of them had said anything to the other, Bennett should have asked the man what he meant. But instead, he paused for a moment, looked at him, and then continued to step into the water. After all, to Bennett's knowledge the water had always been clean, and, as huge as the bay was, presumably nothing could change that.

While the old man watched, a small wave quickly rose above Bennett's knees and then washed back down and across his toes, the sand sliding out from underneath and tickling him for a moment. He kept walking, a swaying, unsteady maneuver, until the water reached his chest, and then he bent down a little to let a small wave temporarily submerge him. He began swimming, counting his strokes. Four hundred strokes equaled half a mile, and he had time for at least that. As relaxing as his swim was, he couldn't loaf in the waves. He had to finish, shower, dress, eat breakfast, and be in class by eight o'clock. Perhaps that's why he hadn't replied to the man—his time was limited.

As he left the surf line, he watched the blurry breakwater to his right. The row of rocks jutting out from the beach was designed to reduce sand erosion, but the barnacle-encrusted obstacle could also slice a swimmer into shark bait if the current pushed him into it. Bennett had been thrown into the rocks before. Clinging to one, he had slowly, painfully crawled out of the water while the waves pulled at his body as if trying to saw him in half. He had scraped himself over the sharp barnacles so badly that he bled from chest to feet.

Once he made it past the breakwater, Bennett turned left, to the south. He watched the shore for landmarks, not so much to measure his distance, given that he counted his strokes, but to make sure the currents didn't pull him farther from shore. But on this morning, out beyond the surf line, he just bobbed up and down in calm water while the as-yet-unformed breakers rolled through the water under him. There was little sound out here. Occasionally, birds flew by, some so low he could hear their wings beat against the air, and rarely, he would see an otter or even a dolphin. As the years went by, he had worried more and more about sharks, as though the chance of an attack got higher the longer he spent in the ocean.

Bennett didn't think much of the man's warning, although the natural storyteller in him was already processing the moment, hoping to shape it with a lengthy prologue, crafted details, and finally, after all the buildup, a denouement (prefaced by "to make a long story short"): "He waved this newspaper in the air and said . . ."

If asked about the warning itself, he might have said he realized that, although he once could see the ocean bottom as he swam, now it was murky, a gray translucence. Later, he would tell people he occasionally tasted something funny in the water, but, really, that was just his storytelling again. While he swam, he actually tasted nothing, not even the salt water. And generally, the water's turbidity so close to shore stirred up the sand and prevented the kind of clarity he claimed he'd seen.

But these were all momentary musings, the kind that might occupy others for longer. Bennett's mind simply wandered from idea to idea as if he were trying to clean out his mental files. He thought about his job, his wife, his son, or, if he allowed the memory to intrude, his daughter, who had died a year before when a truck hit her. And in the background, the strokes kept adding up, counted off almost subconsciously. With every other stroke he turned his head to the right and sucked in a calm, relaxing breath. True, the cold water was already drawing away his body heat, but he didn't feel it yet. The swimming itself was automatic, soothing, a meditation.

In about ten minutes, he reached 200 strokes and turned around, heading north, back toward the breakwater and the old man still fishing. He counted out another 150 strokes. He knew from experience that fifty more would probably get him to shore, so he pointed his body toward the beach. Soon, with the waves pushing him forward, his hand slapped sand and he stopped. Four hundred strokes: half a mile.

Bennett dropped his knees into the sand to steady himself, and, after a small wave patted him on the back, he quickly stood so another wave couldn't knock him over. You can't trust the ocean, he always told himself. It sneaks up on you.

His body violently shook as his wet skin hit the winter air. His core temperature had no doubt plummeted; he was nearly hypothermic. Every morning, it was the same—it could take two, three hours to fully recover. He might be standing before his second-period class before he realized he was no longer shivering.

He passed the man, who looked at him again, this time with an incredulous stare. "Water poison!" he repeated and waved a newspaper at Bennett. Bennett couldn't read the *Los Angeles Times* headline, but it said, "Report Confirms Toxic Dumping; Hayden Decries Damage to Bay."

The marine life was dying. Chemicals had poisoned the water. Once he learned the details, Bennett had the most dramatic storyline of his life: He had been swimming in dirty water.

CHAPTER 2

The Witness

Nothing trumps a three-eyed croaker. Dr. Rimmon Fay added the triple-eyed peeper to the freak show of fish he had pulled from Santa Monica Bay as proof that the waters off Southern California were disastrously polluted. In the jar of horrors that he took to various government hearings on the bay during the early 1980s, he showed off fish with other, more common afflictions too—cancerous black tumors, deformed spines, and fin rot—all caused, he said, by the barely treated sewage pouring into the bay every day. The pickled white croakers stared at the unmoved officials, who invariably figured these pathetic corpses didn't speak for the bay's overall health, which they thought was pretty good.

Fay's beloved bay was barely alive, its marine life, in his mind, decimated by Los Angeles' inescapable need to dispose of everything its 3.2 million citizens flushed down their toilets. And, for that matter, because of the industrial waste—including DDT, PCBs, metals, and other toxicants—that went into the same system. Eleven percent of the city's wastewater came from industry. Oh, and let's not forget the storm runoff that drained directly into the surf after an infrequent rain had washed the streets clean of every possible urban pollutant, from motor oil to dog feces. It all ended up in Santa Monica Bay.

In 1985, with the exception of the storm drain runoff, this flood amounted to 420 million gallons a day flowing to the city's Hyperion Sewage Treatment Plant, which sent all but the evocatively termed "solids" through a five-mile-long pipe into the bay. Occasionally, the stream of raw sewage—its volume equal to the state's tenth largest river—would burst out of a pipe or slosh from an overflow device before reaching Hyperion, and pour down the nearby concrete-lined Ballona Creek and directly into the surf. (The so-called creek is no more than a large, trough-shaped canyon running through the city and ending at the ocean; it was designed to prevent flooding during the area's brief but stormy rainy season.)

None of this was obvious to the people who swam, surfed, or sunbathed at any of the eighteen beaches along the bay. Diseased fish, bacteria counts, and sterile crustaceans weren't exactly apparent from a beach blanket's perspective. Fay, with his two doctorate degrees in biochemistry and chemical oceanography, tried to educate the unenlightened, but discovered that he generally reached only a few environmentalists, whose abilities to spread the gospel according to Rim were no more skillful than his own. Trouble was, even with their occasional hyperbolic announcements that people were swimming in crap— announcements often printed in local newspapers—somehow the outrage that seemed so predictable never appeared.

Fay was not an awkward man in front of crowds, reporters, or cameras, but he explained the bay's fetid condition in a grumbling, academic manner, occasionally slipping in scientific vernacular such as *benthos* or *water column*, as though, for all his experience and knowledge, he couldn't find the kind of incendiary language that would incite anger or disgust in others. He mainly abused decision makers at sparsely attended public meetings, as if he figured the people he believed had caused the problem could, with helpings of dyspeptic speeches and science lectures, be convinced to solve it.

With a certain logic that often went unrecognized, he positioned himself as a witness. Over the course of fifty years, he had watched Santa Monica Bay go from being one of the world's richest fisheries in the 1930s and 1940s to being a place largely devoid of the huge sardine and anchovy schools that once supported further-up-the-food-chain striped marlin, blue marlin, Pacific marlin, yellow tail, blue fin, and barracuda. As a five-year-old kid in 1934, he had watched older boys dive off Venice Pier into water so clear they could snatch the pennies tourists tossed there for good luck. Patrick Wall was among those to whom Fay told this story. "And he says, 'Now you can only see five feet down if you're lucky.' But he was never sad about it. I don't think he took the time to let the sadness sap his energy. He was putting his energy into trying to make it better," Wall says.

Fay dove with scuba gear for hours each day, collecting specimens, most often sea urchins, that he sold to researchers throughout the United States who were studying, for example, human nerve cell damage, tumor prevention, and nonaddictive pain relievers. He also gathered sea urchins for teaching kits that he called the Beginning of Life, which he assembled for high schools and sold for about fifty dollars. An urchin would be injected with a chemical that caused it to shed its gametes.

"You'd look under a microscope and you could see strings of eggs and sperm coming out of it," his friend Don May says now. "You could see the sperm attack the eggs, and the fertilization membrane forms and the cell divides and life starts right before your eyes." The schools and researchers put in orders for the specimens or kits, and Fay took his aluminum boat, the *Torpedo*, to locations on the amorphous water that he knew without a map and then plunged alone to the bottom to gather the animals. When he ran out of air in one tank, he'd surface, swim to the boat, and put on another. He would go through five or six tanks in one day before going back to his Pacific Bio-Marine lab in Venice to prepare the specimens for shipping. It was hard to believe a man could be more at home in the ocean than Rim Fay. He could fall asleep in the murky water.

Unfortunately, the outrage he felt about the environmental deterioration he had witnessed underwater made him a pest in the eyes of decision makers. They watched him approach the microphone and heard this sandy-haired, middle-aged man, with the belly of a guy who drank too much beer, relentlessly snarl at them and their policies. No matter how academic he could be in his recitations of the facts as he believed them, they nevertheless felt attacked. He simply came across to them as cantankerous and uncompromising.

He *was* cantankerous and uncompromising. However, those who nearly worshipped him—he had few friends but many, many admirers—saw that as a lovable flaw. There was something almost charming about his angry sincerity. It was even seductive. "He was a real ladies' man when he was younger," Wall says. "Very suave. And he always knew what to do with a smile and a word." Once, while working as a marine biology consultant for the 1982 film *Cannery Row*, Fay called his best friend, Don May, and told him to quickly get the special at Wild Flour Pizza in nearby Venice and bring it to the MGM studios. Seems he had invited the actress Raquel Welch to lunch, and the three of them sat at the end of a movie-set pier, possibly talking about three-eyed croakers.

Naturally, those who made the decisions figured a little irascibility went a long way, and they generally dismissed his case. *Whatever that might have been. Something about dead fish?* Still, for years, he harassed them, testifying, writing opinion pieces and articles, and giving quotes whenever reporters needed an educated opinion on just how bad things were in order to counterbalance official, rosier versions of Santa Monica Bay's pollution levels.

"I saw the bottom rotting out," he told Richard O'Reilly of the *Los Angeles Times* in 1982. "I saw the animals dying. It was a brutal experience."

And so Fay's fight continued on the afternoon of Monday, March 25, 1985, three days before Howard Bennett would hear the old man's warning on the beach. The fifty-six-year-old scientist took his sideshow of pickled white croakers to a small auditorium in downtown Los Angeles where California's Los Angeles Regional Water Quality Control Board was meeting to consider a seemingly arcane section of the 1972 Clean Water Act, called a 301(h) waiver, which permitted the applicant—in this case, Los Angeles—to sidestep certain federal sewage treatment requirements.

The waiver's regulations took up 10 pages out of the 234-page act and might have made scintillating reading for a sanitation engineer, but few people, including environmentalists, even knew the waiver existed. Certainly on this day, only five enviros spoke against Los Angeles getting the waiver.

In some ways, a 301(h) waiver is fairly simple to understand. The Clean Water Act—which was written to protect fish, shell-fish, and wildlife, not humans—requires that all municipalities treat their sewage in two ways. First, primary treatment removes solids in effluent by slowly moving wastewater through tanks for an hour or two, where anything that either sinks to the bottom or floats to the top is taken away. To be sure, this isn't just the obvious wastes we flush down the toilet. It also includes the fats and oils American diets are known for, sand, and so-called floating materials, such as rags and feminine hygiene products. Then secondary treatment comes along to largely finish the job, using bacteria and protozoa to munch away on the sewage's dissolved biological content. Bonus points go to the rare cities that take this a step further, to tertiary treatment, which cleans the remaining water so thoroughly that it's safely drinkable, although it's been said a glass of the stuff still has the kind of bathroom bouquet no one wants coming out of the tap. (Instead, it's often used to water golf courses or agricultural fields.)

Coastal cities at the time of this 1985 hearing believed mere primary treatment worked fine because they dumped the results into the ocean, a place so vast, they argued, that the wastewater was harmlessly diluted. Los Angeles estimated its dilution rate was eighty-four parts seawater to one part wastewater. Don't make us spend millions on secondary treatment, they told the Environmental Protection Agency, which is in charge of the Clean Water Act, if it isn't even necessary to keep the ocean clean.

With that in mind, the Los Angeles County Sanitation District, which ran the huge Carson Sewage Treatment Plant south of the City of Los Angeles—among other coastal dischargers—spurred the campaign for an exemption from the law. Lucky for them the original law had created a National Commission on Water Quality to evaluate the act for any unnecessarily draconian requirements and recommend changes. Even more fortuitously, Charles Carry, then the district's head of technical services, had been chosen to provide input to the commission. Apparently, he made a persuasive case for backing off on full secondary treatment in certain cases, arguing the common notion at the time that *dilution is the solution to pollution*. As Robert Miele, who worked under Carry at the time, recalls, "Because we discharged into deep ocean waters, which was a great deal of initial dilution, we didn't feel it was appropriate to have to go to secondary treatment for our ocean districts."

Congress took the commission's 1976 report and revised the law in December 1977 to include the section 301(h) waiver. However, the waiver wasn't a free pass. It still required municipalities to meet eight criteria that theoretically ensured a healthy environment. These included protection and propagation of a balanced indigenous population of fish, shellfish, and wildlife. The BIP, as waiver aficionados called it, was a crucial idea, given that the large volumes of organic matter that spewed from primary-treated sewage could reduce the variety of species living in an area by consuming the oxygen in the water and leaving little for anything else. That is, only the few animals adapted to this type of environment hunkered down in the silty muck, and the rest migrated to cleaner waters or simply died. This made for anything but a balanced indigenous population.

And this was Rim Fay's biggest beef. He claimed he had seen species diversity dwindle to such an extent that in some places only a few organisms remained. "Heavy metals, copper, mercury, cadmium, lead, arsenic and even cyanide that are illegally dumped in Los Angeles sewers find their ultimate release at the Hyperion outflow," he told the *Los Angeles Herald Examiner* in 1985. "And because of this the area has seen the disappearance of many unprotected forms of marine life I consider very important. Various forms of crustaceans have completely disappeared. The tunicates [plantlike animals that usually attach to rocks] are definitely inhibited. The diversity of sponges is low, and the most significant reduction in diversity is among seaweed and algae."

The City of Los Angeles' Hyperion Sewage Treatment Plant had once been a full secondary-treatment facility, starting in 1951, when the city—which had been quarantining the beaches immediately north and south of the Hyperion because of all the sewage-related bacteria in the water—rebuilt the facility at a cost of forty-one million dollars. But even then, according to Frank Flood, a consultant hired by the city to evaluate Hyperion, the plant was already at 90 percent capacity. "The city had better start plans now or I predict that Hyperion will be overloaded before anything can be done about it," Flood said. And he was right. By 1958, the plant—designed to take in 100 million gallons a day—couldn't handle everything the growing city of 3.2 million people flushed its way and resorted to discharging some wastewater after giving it only primary treatment. By 1985, the aging plant—first built in 1894 as simply a central location to dump raw sewage into the bay—was providing secondary treatment to just 25 percent of the 420 million gallons it processed daily.

If it sounds as though the city had at least tried to be virtuous by adding full secondary treatment, keep in mind that it also dumped the sludge—the solids remaining after primary treatment—out a seven-mile pipe built in 1957. To people like Fay, this made no sense. The city, he said, basically cleaned up the wastewater so it presumably wouldn't hurt marine life, but then spewed into the bay the very stuff it cleaned out of the raw sewage to begin with. Congress realized the same thing and amended the 301(h) waiver in 1981 to disallow sludge discharge into the ocean. Just before that, the city had missed an April 1980 court-ordered deadline to put the sludge through a largely federally funded high-tech incineration process that would yield electrical power to run the plant, and was now staring at a July 1985 deadline it already knew it couldn't meet, either. In fact, Los Angeles had a fairly long history of promising to stop sludge discharge. As early as 1974, it had said it would collect the sludge for incineration the following year but instead continued dumping it.

In the meantime, Los Angeles applied for the 301(h) waiver in September 1979, two years after it became law. In 1981, an Environmental Protection Agency task force made up of scientists and staff members tentatively approved the city's waiver. From that point on, Los Angeles simply continued doing what it did, waiting for the waiver to be officially granted. At the same time, a previous federal lawsuit to force the city into full secondary treatment was put on hold. Owing to a number of factors, the public hearing on that waiver was scheduled for three years later, on March 25, 1985.

But this is where the story gets a little murky. Los Angeles had to apply for an NPDES (National Pollutant Discharge Elimination System) permit to operate Hyperion. The state's Los Angeles Regional Water Quality Control Board, which issued the NPDES permit, could, if it wanted to, advise the EPA to not grant the 301(h) waiver as part of the permit if the board believed that water given only primary treatment would cause environmental damage to Santa Monica Bay. In addition, the city had to follow a 1972 state law called the Ocean Plan, which didn't require full secondary treatment but contained its own set of environmental protections.

As the EPA's Patricia Eklund reminded those at the March 25 meeting, the EPA would deny the waiver if the water board so recommended. According to the board's former executive officer, Robert Ghirelli, the board figured that, if the EPA had tentatively approved the waiver, they weren't going to disagree. In fact, the water board did little to study the matter themselves, depending instead on the EPA's own research conducted between 1982 and 1985. To Rim Fay this meant the fix was in and the hearing that day was a mere formality.

In a way, that it was a mere formality applied as well to a legally required notice for the meeting, which had been placed about a month before in the *Los Angeles Times'* classified section and referred to the NPDES permit modified to include the waiver (at the meeting, Eklund introduced a copy of the notice as exhibit A prior to the testimony). But unless readers knew what *NPDES* meant (provided they even saw the announcement), the sewage jargon would have been lost on them. And if anyone bothered to read that notice, they would have had to go to the water board's downtown Los Angeles office to read the meeting's planned agenda. No doubt for these reasons, the 301(h) waiver attracted only a few people to the Monday afternoon meeting. To an outsider, it probably seemed as though the EPA and water board just wanted to get the waiver process over with.

To that end, the five government representatives who first spoke treated the board's endorsement of the waiver with a kind of certainty, concentrating instead on the water quality monitoring program required during the permit's five-year period. No one actually asked the board to *approve* the waiver.

Harry Sizemore, assistant director for the Los Angeles Bureau of Sanitation, got so far ahead in the process that he wondered out loud if the city could slack off on the monitoring after a while to save a portion of the $3.5 million he estimated it would cost. "The city is most concerned about the size and scope of the ocean monitoring . . . ," he said. "Our hope is that after a period of time you will be able to review our efforts and will eliminate perhaps some of the non-productive expenses."

Following the pro-waiver contingent, Fay's friend Don May, a large, effusive man with a helmet of curly hair, approached the board as a representative of Friends of the Earth, a group he helped form and one of the many enviro organizations he had spent time with over the years. While Fay and another friend, Martin Byhower, listened, fifty-one-year-old May reviewed his bread-and-butter argument—that Santa Monica Bay was a world-class fishery early in the century but had faded as pollution increased. "The only thing that's changed is the permit number," he groaned. "The impact on the ocean just gets worse and worse."

Board member Betty Werthman showed little enthusiasm for the secondary-treatment construction costs May was pushing on the city—"money," she said, "we don't have right now. We have to look for a short-term solution, and this waiver happens to be one short-term solution." With the EPA pulling back on grants to pay for secondary treatment systems, she worried that the public—paying higher sewer bills—would feel abused by footing the tab to solve an issue they "didn't even participate in."

Fay followed this typical argument over money. He was medium height and barrel chested, with thick hands and a ruddy but handsome face. He looked at the board with a familiarity that came partly from his once being a decision maker himself, working on California's first Coastal Commission in 1972 (spawned by the Ocean Plan) until the other commissioners got tired of him railing against seemingly anything that involved development, and he was dismissed.

Fay liked to call himself a simple fisherman, and when his turn at the microphone came, he calmly outlined his fifty-year relationship with the bay, implying ownership or even marriage. "I've probably spent more time under and on the water in Santa Monica Bay than any person in the United States," he said, implying that, since his observations spanned so much time, they couldn't be discounted.

Then his civility wilted as he quickly cut to the chase—the bay's thorough "demise," as he called it. "There's no other answer," he thundered. "Clean water means more treatment, not less!" The left side of his mouth unintentionally twitched upward, looking like a sneer. Once he had everyone's attention, Fay continued his confrontation, banging out each word with slow, growling emphasis. "Clearly, primary treatment has not worked! . . . There is nothing to indicate primary treatment has been adequate. Nothing! In fact, the record is quite the contrary."

As if he were trying to alienate the board even further, he switched to a lecturing tone for a moment's recitation on why secondary treatment worked, and then went on to one of his favorite topics, the loss of species diversity. Finally he said, "Self-monitoring itself is inherently questionable," implying that the dischargers in the house might fudge the numbers in their favor. "Secondary treatment is itself self-monitoring. It's a biological process."

After twenty minutes, Fay suddenly quieted, as though so disgusted he just wanted to leave. He grumbled that he would send them his written statement by the end of the week and abruptly left the podium.

It was an odd, troubling performance. He sounded tired, seemingly straining to maintain his pique. Nothing he said was new. It came out of his greatest-hits bag, one brief comment or complaint after another, strung together with the only connective tissue he had, his anger. And even that appeared blunted by his frustration. He felt ineffective and looked forward to going back to his lab and the six-pack of beer he would finish by himself. The waiver, he assumed, would be granted. The bay's pollution would continue.

CHAPTER 3

The Coalition

For the sake of a good story, Howard Bennett rarely worries if his tales don't always portray him in the best light. Ask him if he knew anything about Santa Monica Bay's pollution before the old man's warning and he'll tell you a seemingly unrelated tale of how, after a rainstorm, he and a lifeguard friend swam north from Playa del Rey beach to where the nearby swollen Ballona Creek emptied into the ocean. The current rushed through so strongly there that it was like heading into an aquatic treadmill. They stroked and stroked as hard as they could just to stay in place, pumping against the force with all the masculine pleasure that comes from defying such power. Unfortunately, they didn't realize at the time that all the city's storm drains that poured into the creek's concrete channel were filled with the worst kind of muck the streets and drains had to offer—engine oil, fecal material, chemicals. The runoff was so polluted that the turbulent water turned a soupy brown. And they joyfully swam in it.

So, given his ignorance of the situation (one shared by the rest of the city), when Bennett heard the old man call the bay poisoned, the possibility that he had been swimming in anything but pristine water barely registered. About the only thing that suggested the man's warning was more than crazy ranting was the fact that his reference material was the *Los Angeles Times*. Even so, Bennett wasn't worried enough to pick up a copy that morning and read the article for himself.

Instead, he stuck to his usual routine, which was calculated to get him from his daily swim to his classroom with the kind of punctuality he had obsessed over since his boarding school days in New Hampshire, some four decades before. Get up at six. Start swimming fifteen minutes later. Swim no more than twenty minutes.

He then ran across the beach back to his house, shivering so violently that he could barely direct his body in a straight line. He immediately took two hot showers, the first under a nozzle Bente had installed outside the first-floor apartment, and then, after he drained the hot water heater there, he took another shower in his bathroom, using up its hot water supply as well. With the cold still lingering in his body, he put on a suit and tie, ate the usual bowl of cereal Bente put out for him, and left the house no later than 7:30. He drove twenty minutes to Culver City High School, parked, signed in, and entered his class just before eight, his English literature students unaware that tiny spasms yanked at his body until he finally warmed up by second period.

With just enough self-deprecation to make anyone believe he was telling the truth, he claims he chose English because he figured it was the easiest subject to teach (his math skills barely enabled him to calculate his age). With seventeen years in the classroom—and after realizing, to his dismay, that English teachers grade mountains of homework—Bennett taught Shakespeare, science fiction and fantasy, tenth grade literature, and myth and legend.

And this is what preoccupied him throughout that day. He didn't think about the old man, nor did he ask others if they had seen the newspaper. The news itself didn't ripple at all through the teachers' lounge or the student body, even though many of the students were surfers who spent more time in the ocean than Bennett did. No one mentioned the article's main point, that the toxic chemicals—DDT, polychlorinated biphenyls (PCBs), cyanide—had been dumped in the bay for years. At just one dump site off Santa Catalina Island, forty miles from Bennett's house, about 770 tons of DDT had been *legally* spilled into the ocean, presumably a place so big that, according to common wisdom at the time, there was plenty of room for even a few poisons.

It's not that this was breaking news. During a week or two the year before, reporters, one or two politicians, and some environmentalists had all been wringing their hands over the same details when they first surfaced, but, as appalled as the public might have been, interest died away. That's the thing about DDT buried several hundred feet underwater, where you can't see it. Getting your kids to school and paying the bills always take precedence. (The same information had come out earlier, in 1970, but again, any alarm it might have caused quickly dissipated, much as the DDT was no doubt expected to do but didn't.)

Just the same, if someone yells, "Water poison!" it makes one at least a little curious, so when Bennett returned home that

Howard Bennett working the phones in his living room in 1985.
Photo courtesy of Bente Bennett.

afternoon, he phoned the only person he knew with a scientific background—a part-time lifeguard and full-time marine biologist named Rim Fay—to ask him what, if anything, was happening.

Bennett had met Fay when another lifeguard, Bud Williams, invited Bennett to ride with him to Santa Monica, where they would drop off a lifeguard truck and then swim back to Playa del Rey. Given that the distance was about three miles, this was nearly as crazy as swimming in storm runoff. In Santa Monica, near the pier, they met Fay, who recognized the pair's overactive testosterone at work and offered to cut the distance in half by driving them to Marina del Rey, where they could swim through a boat channel at the marina, out to the ocean, and then to Playa del Rey. During the ride, Bennett learned Fay spent most of his time working as a marine biologist. Water quality never came up in the conversation.

If nothing else, this established Fay in Bennett's mind as a walking resource for marine life information, and so that afternoon, on March 28, with no sense of urgency, Bennett rang Fay at his Venice lab and casually asked him about the article. Still seething over the previous Monday's hearing, Fay sidestepped the DDT and PCB issue and bitterly explained that the city, in his mind, was about to win the 301(h) waiver. In his deep growl, Fay told Bennett about the Clean Water Act and how it mandated full secondary sewage treatment unless Los Angeles could secure that waiver. He correctly surmised that the Regional Water Quality Control Board—a state agency Bennett had never heard of—had already concluded that the EPA's prior approval of the waiver was good enough for them, and that they would concur with its decision. Why was this? Because so few people had attended the meeting, Fay said. As far as the board was concerned, a few fanatic environmentalists weren't enough to convince them there was anything wrong with dumping partially treated sewage into the bay. With that, the opportunity to kill the waiver had passed.

Fay implied there was something conspiratorial in all this, while knowing there probably wasn't. After all, the water board rarely saw public attendance beyond a few people such as Fay, and they just assumed the lack of involvement reflected widespread apathy toward the esoteric topics they typically covered. Certainly, they weren't trying to hide the 301(h) waiver issue from the public. Rather, they didn't think anyone cared.

But Bennett—with Fay's coaxing—saw it differently. The city wanted this waiver—it *needed* the waiver to save the construction costs involved with building a plant for full secondary treatment. In Bennett's mind, the board had prevented public protest by purposely burying the meeting announcement deep within the newspaper. And so with the kind of logic that comes from fresh-faced naivete, Bennett told Fay, "We have to force the EPA and water board to hold another hearing!"

"There's nothing you can do about it!" Fay snapped.

"They have to hold another hearing!" Bennett repeated, so sure he was right. "We're going to change this. We'll make them do it."

Bennett has told this story a hundred times, always with the same indignation, as though the visceral reaction he says he felt back then remains, still cooking, inside. Even if it seems obvious that this was a personal issue for him, that he was disgusted with the idea that he had been swimming all those years in what appeared to be tainted water, he claims the anger he felt at that moment had nothing to do with him. This was about *justice*, about the public—not some schoolteacher—being treated with such disregard that government officials would actually allow them to play in an ocean loaded down with sewage. If the citizenry knew about the 301(h) waiver, he optimistically assumed, they could force the EPA, state, county, and city to clean up the bay.

It's hard not to inflate this moment into a pivotal point for the environmental movement in the country's second-largest city. Fay had unwittingly provoked Bennett with the irresistible combination of an apparent injustice that affected not only him but also thousands of others and the simple challenge that comes from saying, *You can't do that.* It seemed the usual dull tactics Fay and others had employed were about to be tossed aside in favor of brash, loud demonstrations by Bennett-led individuals intent on getting the kind of publicity the cause had yet to see.

Bennett was a self-described bulldog (others have twisted the canine metaphor in another direction and called him a publicity hound) who believed relentless press coverage was the best method to change the world. Being told he couldn't stop the 301(h) waiver from going through gave Bennett a compelling reason to abandon his secure, mild world and gnaw on a few politicians' ankles.

He had done this ten years before in a smaller fight involving a proposal by the South Coast Regional Coastline Commission to convert private coastal property near Bennett's home to public-use land through eminent domain. With a natural talent for acidic rhetoric, Bennett boiled down the issue to the sort of emotional terms that moved crowds to anger. In this case, he and a neighbor, Ruth Lansford, named their fight the Brown

Spot Campaign after a commission member pointed at a map and indifferently suggested the commission include in their plan a twelve-acre brown patch where dozens of homes stood. Protesters led by the two showed up with brown paper cutouts symbolizing the people affected, along with signs saying, "Confiscation Without Representation," and the shocked commission quickly backed down.

There were other campaigns, as Bennett called them. His activist résumé included numerous protests, beach cleanups, and fund-raisers for charitable causes, each accomplished so rapidly that he believed his ability to garner publicity and sell any humanitarian idea was nearly infallible. All it took was for him to deem that an injustice had been perpetrated, and his anger propelled him forward.

Nevertheless, Fay couldn't believe anyone would be so stupid. Who was this idiot Bennett who didn't know what it was like to trudge from meeting to meeting, berating guys like the members of the water board for years and getting nowhere? Bennett hadn't spent hours and hours in uncomfortable chairs waiting for his few minutes to plead for the bay's health. He hadn't written editorials and articles or spoken to endless reporters all looking for a quick quote or sound bite that amounted to nothing more than a brief, impotent rebuttal of some public official's reckless belief that "dilution was the solution to pollution." Bennett didn't dive all day among the dwindling sea critters, studying them, collecting them, handling them like children. All he knew about the ocean was what it was like swimming on the surface a few feet from the beach, never seeing the tumors and fin rot. No, this Bennett guy—someone Fay hardly knew—clearly couldn't understand. In the kind of dismissive pique he had shown others, Fay yelled, "You're crazy!" and hung up.

Bennett's a little vague on his next step, but it's clear he didn't leap right into the waiver issue. Like most of us, he found the path of private complaint the most comfortable at first. Standing there in his living room, watching Santa Monica Bay's compromised waters now black and cold, the crescent moon about to set, he filled the evening with a rant to Bente, his only audience at the time, about justice. That's right, he said, justice. He took this far beyond unnamed bureaucrats plodding through their jobs. This involved right and wrong. You don't hold a public hearing and not tell anyone about it, he barked at his wife. You don't pollute the water! As angry as he felt about the government's apparent deceptions, he also zeroed in on the personal element. He and his family had been swimming in this filthy water for years and no one had bothered to tell him it might be filled with shit! Excuse my French, he added.

Bente tolerated his indignation and occasional profanity long enough to get the gist of his anger. However, this wasn't a woman who believed jawboning about the problem did the world any good—and, indeed, she saw situations such as this globally, usually before her husband did. Sure, what Howie said angered her. She believed in the government's benevolent duty to handle such issues in ways that benefited the public as well as the environment, and she assumed the city was doing the right thing with regard to sewage treatment. Bente didn't think about it every time she flushed a toilet, but then again she shouldn't have had to. As with the electricity dependably feeding through her house or, more to the point, the clean water coming out of its taps, there was nothing wrong with taking for granted that the city would do its job. She loved the ocean just as much as Howie did, and even though she didn't go swimming in it as often as he did—because she didn't like the cold water—she figured it must be clean, because that was the government's responsibility.

As Howie cursed the people who allowed the pollution to continue, Bente's trust in the government was being crushed. His tirade shouldn't be the end of it, she thought. After all, she had been born in Denmark in 1925, and when Nazi Germany invaded the country in 1940, she responded by joining the underground army. Hardly the reaction of someone content to quietly complain, nor that of a typical fifteen-year-old. She had once delivered ammunition in a baby carriage by wheeling it past German soldiers, and if she could do something like that, then challenging the City of Los Angeles wouldn't be difficult. If nothing else, no one would get killed.

Nevertheless, Bente was a shy woman who didn't have Howie's comfort in front of audiences or his storyteller's talent for riveting attention on him and his message, usually in that order. So as soon as he exhausted the subject as far as he knew it, she told him that, if he was so incensed over the pollution and the waiver, he had to do something about it. Her underlying message was clear: I'm as angry as you are, and I want you to fight it! In a way, the subsequent campaign was a useful combination of his indignation and her conscience. Things had been that way since they had married thirty years before.

Howard and Bente had met in early 1954, when the Royal Danish Polio Foundation had sent Bente, a physical therapist, to New York's Rusk Rehabilitation Center at Bellevue Hospital to research alternative therapies for polio patients trapped in tank respirators, known as iron lungs, which kept them alive. At the time, Bennett was driving a cab, and he picked her up along with a wealthy quadriplegic patient Bente was working with on the side. According to Bennett, who preferred wisecrack explanations to any deep self-analysis, love came with the first handshake. Bente had nearly crushed his hand. The twenty-four-year-old cabby who was still searching for a career had found his future wife, the twenty-eight-year-old woman with hands as strong as an ironsmith's.

Bente's reaction had been more subdued. Bennett was just a good-looking, well-behaved boy. That boy—after a courtship in New York City—followed her to San Francisco later that year, when Bente got a job at the Cabot Kaiser Institute. They drove cross-country together in a new car Bennett had volunteered to deliver for a friend at a car dealership. They married that November in Santa Rosa, California, getting there in a used sixty-five-dollar DeSoto.

Bente taught Howie about compassion. She would invite polio patients to their home for dinner, asking her new husband to pull the wheelchair-bound guests up the three flights of stairs to the couple's apartment. To raise a teenage patient's morale, they took him to a rodeo—he had been raised on a ranch—and had him do the play-by-play for them.

They kissed in front of friends. Howie publicly proclaimed his love whenever he could. They grew closer and closer over the years. In 1961, after moving to Playa del Rey, they adopted Leif, who was just six days old. The next year they adopted month-old Danya. The family swam in the ocean that was just outside their back door. They traveled around the Southwest and later around the world. Howie became a teacher in 1967 after working as an appliance salesman (this was almost a matter of familial continuity; his father had been a traveling salesman, his mother a teacher in New York City's Hell's Kitchen).

For Howie, this life—happier than most, it seemed—could not have been possible without Bente. This slight but strong, attractive blonde, now a physical therapist working for a doctor in Beverly Hills, dazzled him every day. Her gray eyes pierced him, stilling his temper, while her cool Danish logic provided him moments of clarity. The articulate storyteller in him limped away whenever he talked about her, unable to express himself without resorting to hackneyed phrases. "I don't know if I could live without Bente," he told people, miserable at the idea that his beloved wife could be anywhere but in that boxy gray house on the beach.

And so when Bente strongly suggested he pursue that second waiver hearing, Howie the bulldog blindly charged forward, growling at every obstacle, eager to please his wife, someone who truly knew about justice. But the effort was going to require more than one schoolteacher with a talent for gab. So Bennett contacted his neighbor Ruth Lansford, a slender, gregarious woman who had lived on the beach about as long as the Bennetts and who was a cofounder of the Friends of Ballona Wetlands, an environmental group. Her organization was currently deep into trying to protect the nearby wetlands, a small but vital remainder of what had once been many wetlands along the California coast. The Playa Vista Corporation wanted to develop the wetlands into a housing project the size of a small town, and the Friends had challenged it over and over in an effort to stop the bulldozers.

Because of her background, Lansford owned a bulging Rolodex of people already entrenched in some way in the environmental movement in Southern California. Bennett realized that here was the human foundation for launching his campaign. He needed names, lots and lots of people and organizations who could help him.

It was all about perception. A small bunch of people railing against the 301(h) waiver was nothing more than a gnat in city hall, buzzing but barely heard. However, if a dozen or more groups were all raising the same banner against the waiver, then suddenly the city, the EPA, and the Regional Water Quality Control Board would have to listen. Or so it seemed to Bennett, who believed with faith bordering on fanaticism that his goal—the waiver's denial—would quickly, inevitably be realized.

One of the first people he spoke to was Janet Bridgers, who, along with her husband, Patrick Wall, ran Eco Features Syndicate, a news syndicate devoted to environmental stories. They had in fact sent out articles earlier that year, including one written by Rim Fay detailing the bay's polluted condition. Bridgers suggested Bennett form a coalition and make it clear this was a widespread movement encompassing potentially thousands of people. The rabble was being roused. Or, more to the point, they were voters who didn't like what they saw and wanted it changed.

He spent hours on the phone over the next several days, calling the organizations whose names Lansford had given him and asking them to join the cause. Amid all these conversations, someone told Bennett that a scientist from the EPA had rated the bay's pollution as the world's worst. Bennett doesn't recall who told him this, nor does it appear that he tried to confirm the quote. Additionally, it's not clear if the person who repeated this even accurately relayed the quote to Bennett. The actual statement came from an EPA oceanographer and environmental scientist, Dr. Brian Melzian, who had described Santa Monica Bay to the *Los Angeles Times*, saying, "Nowhere on Earth is as heavily contaminated with DDT and PCBs as Southern California."

Bennett translated this into his campaign's single most damning accusation. "Santa Monica Bay has been called the most polluted bay in the world," he wrote several days later in a letter he sent to those on his phone list. Occasionally, he put this in quotes as if to imply it was a direct statement from an unnamed source, or he attributed it variously to the EPA or EPA scientists. There's no evidence that anyone at the time bothered to question him about the quote, which Bennett usually inserted in the middle of a comment about treated sewage coming from Hyperion. Melzian was actually referring to the chemicals that had been dumped into the bay off boats years before—after the state had issued permits for this—and the tons of DDT that had been washed into the sewer system by the nearby Montrose Chemical plant; the DDT had slipped through the Los Angeles County's sewage treatment plant and into the bay.

Still, Bennett now had a compelling argument about more than just the trust issue of government officials allegedly keeping a public hearing secret. The quote put the pollution on a worldwide scale. Not only did it make the water quality appear so horrid that it beat out what one presumed were worse conditions in, say, third world countries, but it also added a component of shame. Being number one in dirty water was hardly something the city could be proud of.

As more and more groups lent their names to Bennett's coalition, he realized he needed a name for his nascent band of dissenters, and creativity in such matters wasn't one of his specialties. Almost as a throwaway gesture to describe his goals, he finally came up with "Coalition to Stop Dumping Raw Sewage into the Ocean." Along with being an easily forgotten jumble of words, the moniker slightly misstated the issue. Raw sewage was not being dumped into the ocean.

However clumsy it might have been, the name did directly proclaim what he wanted to accomplish and even broadened the task, from addressing just Santa Monica Bay to addressing all the world's oceans. In other words, Bennett quickly surmised the bay wasn't the only place where a city dumped its sewage. For example, the New York and New Jersey municipal sewage treatment plants barged their millions of tons of sludge 122 miles offshore and dumped it beyond the continental shelf. Worse, untreated sewage went into the Mediterranean, and poisonous red tides, which are algae blooms bolstered by sewage-related organic matter in the water, periodically sucked the oxygen from bays all over the world, suffocating fish. Seattle's Elliot Bay was contaminated with copper, lead, arsenic, zinc, cadmium, and PCBs. San Francisco Bay contained many of the same heavy metals.

Sewage and heavy metals had a huge impact on local fisheries. As Rim Fay had pointed out in his testimony before the Regional Water Quality Control Board, the Santa Monica Bay fishing industry was so decimated that about the only fish left to catch were bottom-feeding croakers and the like, which made their way to local Asian fish markets and amateur anglers' dinner tables. (Unfortunately, the fish contained DDT and PCBs, enough to make them a public health hazard.) On the East Coast, red and brown tides clobbered the shellfish industry often enough to make a scallop guy wonder about changing careers.

In a small effort to at least imply that the problem affected more than just Santa Monica Bay, Bennett took to calling our planet "lifeboat Earth." While this sounded like the kind of clichéd phrase you might hear from a screeching enviro, Bennett used it sincerely and gently to paint humanity as riding an ultimately fragile vessel, and to point out that we needed to do something before we sank the planet.

Later, as Bennett learned more about the issue, he wrote a note to himself outlining a strategy with the following goal, partly quoted from the Clean Water Act: "To have a swimmable, fishable ocean that meets the goals of the Clean Water Act—'to restore and maintain the chemical, physical and biological integrity of the nation's water.'" He added a list of more specific objectives he wanted to accomplish:

- Defeat the 301(h) waiver—full secondary treatment
- Oversight hearings by Congress on enforcement
- Congress declare SM Bay a Superfund Site
- Building moratorium in Los Angeles because of overcapacity
- Amend the California Ocean Plan
- Enforcement of Pre-Treatment Program
- Enforce a non-industrial source control program
- Amend the Clean Water Act to remove the 301(h) provision
- Plan to clean up storm runoff
- Developers pay their full share of infrastructure to support their developments

Bennett was thinking big. He was projecting this coalition beyond the waiver, envisioning it as a movement with such influence that the city, county, state, and federal government agencies responsible for the state of the ocean would make it their business to keep it clean and unspoiled. His effort wasn't exactly on a par with Bente wheeling ammunition past Nazi soldiers, but he was following her example as best he could.

Within a few days, about a dozen organizations had joined the coalition. It was time to get the press involved.

CHAPTER 4

Squirp

Let's say hello to propaganda for a moment. For the sake of simplicity and instant persuasion, some enviros at the time initially blurred the distinction between raw and partially treated sewage, dependably evoking visceral reactions from anyone they told that Los Angeles—both city and county—was polluting Santa Monica Bay with, *ew*, sewage. They knew this automatically created in almost everyone's mind the not-so-pleasant picture of turds washing ashore while kids played in the surf. In fact, Bennett developed a favorite line for this image, telling reporters, "When I swam, often I tasted things in the water. I thought it was something I ate. I found out it was something someone else ate."

In truth, the Hyperion plant treated the sewage just enough, as Bennett loved to say in interviews, "to separate out the big chunks," in a process called primary treatment. The raw sewage, after going through screens, was sent to large tanks, where suspended solids either floated to the top or sank to the bottom.

And the result, a still nasty mixture of organic matter, bacteria, and suspended toxic metals, among other things, poured out the five-mile-long, thirteen-foot-diameter pipe at a depth of 197 feet. Near its end, the pipe split into a Y. Portholes spaced every forty-eight feet along the pipe after this point dispersed the wastewater into what was known as the zone of initial dilution. Here, one part treated sewage was diluted by eighty-four parts ocean water. (The leftovers from this process—the sludge—went through a seven-mile-long pipe and accumulated at the outfall at the edge of the underwater Santa Monica Canyon.)

Just the same, these details were hardly as forceful in terms of propaganda as simply saying, as Bennett and others did, that the city daily dumped 420 million gallons of sewage into the bay (actually, Hyperion *treated* 420 million gallons, and about 25 percent of this received secondary treatment as well, the biochemical process that the Clean Water Act required for the entire volume). Similarly, the name Bennett called his band of protesters, the Coalition to Stop Dumping *Raw Sewage* into the Ocean, was calculated to elicit strong disgust. In fairness to Bennett, he dropped the word *raw* after his son, Leif, correctly pointed out it wasn't really pure but partially treated sewage.

However, the propagandist version of the issue could go only so far. Bennett needed a story he could tell. And that required a hero and a villain. A story was especially needed to attract the Fourth Estate, which responded quickly to plotlines of good and evil. However, just a few days into his campaign, all he had for villains were faceless targets—the City of Los Angeles, the city council, the EPA, and the Regional Water Quality Control Board—which were just institutions grinding away in the land of gray, and too slippery to strictly categorize as bad guys.

He hammered away at them as best he could during his count-less phone conversations with potential coalition members. Without a true villain, he ran with the idea that the March 25 meeting was deliberately kept from the public so the 301(h) waiver could go through unscathed. If nothing else, people loved conspiracies.

Bennett didn't realize a villain hid in plain sight. There he was, a tall, handsome man, a former adventurer who knew how to sell his own variety of propaganda. He did it with expert flourish, convincing the decision makers that the bay was doing just fine. In fact, he assured them, the sewage pouring from Hyperion actually benefited marine life. He went so far as to tell people that the waste products of their digestive process—an entire city's, uh, effluent—actually kept fish and other creatures well fed.

This villain—as Bennett and others would come to portray him—was Willard Bascom, who ran the Southern California Coastal Water Research Project, or SCCWRP (which most people pronounced "Squirp"). The sixty-nine-year-old execu-tive director (who looked ten years younger) oversaw a crew of some twenty scientists and other researchers who—under an annual million-dollar-plus contract with five Southern Califor-nia dischargers, including the City of Los Angeles and Los Angeles County—monitored the bay's waters and marine life to see how discharges from their clients' sewage treatment plants affected ocean ecology. The city and county listened to the likeable Bascom because he sounded so credible, a well-educated staff backed him up, and they were paying good money for the information.

On the surface, this didn't sound like such an outrageous re-lationship, given that little was known about ocean water quality at the time and SCCWRP's basic mandate was to learn more about how sewage discharge was affecting the environment. Created in 1969, SCCWRP at first had many supporters, including Rim Fay, who wrote in the *Los Angeles Weekly*, "I suggested at this time that a great amount of information on the dynamics of inshore waters might be gained by coordinating all of the marine-monitoring programs carried out by the waste-discharge agencies. Since each of these agencies was sampling throughout the [Southern California] bight, I reasoned that if their data were to be pooled, and summarized, it would provide continual spatial and temporal information about the complex marine environment of the area. In turn, such information could be used to better understand what effects, if any, the waste discharge was having on the environment."

In addition, the people running the treatment plants needed this research because their state-issued permits to discharge wastewater into the bay were partly based on their ability to prove they were doing little if any harm to the marine wildlife. Fay was "soon surprised and dismayed" to learn those dischargers were actually funding SCCWRP, considering the conflict of interest this implied. He wasn't sure just how honest SCCWRP scientists could be about their findings, should these turn out to be critical of the company's clients, the dischargers. For their part, the dischargers generally assumed that whatever Bascom told them was based on scientific fact.

Working out of a converted Long Beach carpet warehouse on the Pacific Coast Highway, the researchers themselves—many with PhDs or master's degrees—churned out thick reports filled with data collected in field research or from some of the most sensitive instruments available at the time for sniffing out pollutants.

Their work was as thorough and accurate as it could be, unstained by dogma. Other scientists referred to SCCWRP's research, trusting its validity. It seemed as though no one had a reason to dispute what they reported. (And to be sure, the SCCWRP scientists didn't believe Bascom's assertion that sewage made good fish food, though they never said it publicly.)

But Bascom insisted on writing the reports' summaries himself and edited the research sections, which were reviewed by the dischargers' technical staffs. It's not clear if those at city hall ever cracked open anything more than the summaries themselves. According to those who worked at SCCWRP at the time, Bascom cherry-picked data that supported rosier conclusions than the researchers had actually derived. In some cases, this was as simple as replacing volatile words with vague, less troubling terms. When his scientists referred to the nearly lifeless area around the seven-mile-pipe's outfall, where sludge was dumped, as the "dead zone," he changed the wording to "degraded zone." Another area, which basically encompassed a good chunk of Santa Monica Bay, was first termed by the scientists as "degraded," which Bascom then revised to the fuzzy word "changed." As former SCCWRP scientist Dr. Bruce Thompson puts it now, "He was very careful about his terminology. He defined words for us quite often. . . . So he was very conscious of that. I don't know where he came up with a lot of his definitions, but, you know, he was the boss and I was the dumb little postgrad."

In another example, the scientists estimated that biodiversity in the dead zone, uh, degraded zone, had dropped by about 99 percent (hence their original apocalyptic term), leaving, Thompson says, "two or three species of very weird organisms that could handle it." Among these hardy species able to thrive in the sulfide-rich muck was an interesting creature called *Solemya reidi*, or gutless clam, whose gills host a bacteria that oxidizes sulfide, that is, turns it into oxygen. This is significant because the sludge's organic content sucked so much oxygen out of the sediment that most creatures fled the area or simply died. This symbiotic relationship between clam and bacteria allowed both to survive in an otherwise poisonous environment. White croakers also did well in the area, gobbling up worms that lived in the sludge.

Armed with this information, Bascom concluded in public that the biomass—that is, the total number of creatures—in the area had actually increased thanks to the sludge, and that the remaining species happily dined on the sewage. That was true in one respect: the number of organisms had indeed increased. However, most of the diverse marine life that had previously lived there had moved out of town, looking for less contaminated neighborhoods, or died.

Talking to a *Los Angeles Times* reporter in 1982, Bascom said, "It's very hard to show there is any damage [from municipal waste discharges]. What we do see is change. The critters living out there have reorganized themselves."

Then again, as one former SCCWRP scientist, Dr. David Brown, put it, "It was like rats in a landfill. Yes, there's a high biomass there, but it's only one species and it's not the species you necessarily want."

Preferring their work not be misconstrued, the staff complained to their boss that the public had a right to know all the facts. According to Brown, Bascom told them that being too specific or candid was a slippery slope: the public couldn't be trusted not to overreact if given all the information. Or, as in Bennett's case, they would protest the issuance of a 301(h) waiver, and naturally, SCCWRP's clients' ultimate goal was the coveted exemption from having to spend money on secondary treatment.

As dissension rose among the scientists, Bascom reminded them, in an October 1984 meeting, that SCCWRP's "original purpose" was to provide data to the dischargers that could be used "to rebut the unsupported allegations of 'environmentalists' or to answer questions and erratic regulations of the state or EPA."

He called the issue over secondary treatment a "long-term struggle," in which SCCWRP's clients were trying to avoid spending millions of dollars on supposedly unnecessary treatment plant upgrades. Just the same, he said, as if to console them, "we have the freedom to look into any aspect of real or alleged effects to find out the truth of the matter. . . . The reason we have this freedom is that the original sponsors and Consulting Board believed that the facts, whatever they might turn out to be, would demonstrate that the total costs of secondary treatment (environmental, financial, and other) would be much greater than those of sea discharge. It is most important that our statements do not exceed our evidence. When all the information needed on any point is in and thoughtfully digested, there will be plenty of time to tell the public the answers."

In a sense, Bascom enjoyed this contrarian position. Born in Bronxville, New York, in 1916, he had tasted the underground life while working on the Delaware Tunnel (which later supplied water to New York City) and liked it so much that he attended the Colorado School of Mines in preparation for a mining career. But he was expelled in 1942, a few months before graduation, after what he called a "an altercation with the president over my maverick attitude toward school." For the next three years, he bounced from mine to mine in Colorado, working as an engineer, and explained to anyone who asked that his itinerancy was an effort to learn the mining business.

In 1945, after a chance meeting with a civil engineer researching ocean waves for the University of California, Bascom was hired for a navy-funded surf survey and began his oceanography career. For five years, he plunged into Pacific coast breakers with a thirty-foot military amphibious vehicle called a DUKW (pronounced "duck") and studied wave dynamics and how they were affected by terrain and other factors.

During the Waves Project, he met Rhoda Nergaard, a slender, blue-eyed blonde, in Astoria, Oregon. They married, and she followed him during his research travels, along the way giving birth to a daughter and son.

Bascom went on to work for the Scripps Institution of Oceanography, where his iconoclastic personality was further encouraged by a sign posted above the desk of the acting director, Dr. Roger Revelle, that read, "Fan the flames of controversy." After studying the Pacific Ocean's floor for evidence of plate tectonics, still a theory at the time, he left Scripps in 1954 and joined the National Academy of Sciences. There he directed a project to drill the seafloor off the Mexican coast 11,700 feet below the water's surface. Since this established a world record, he wrote a book about the feat called *A Hole in the Bottom of the Sea*. After that, he began his own business drilling for diamonds underwater and exploring for shipwrecks.

And finally, in 1973, he joined SCCWRP when John Isaacs, the engineer who had gotten him started in oceanography and was now on the organization's advisory board, invited him to take the executive director job. According to Bascom's memoir, Isaacs told him SCCWRP's purpose was "to remain neutral while obtaining scientific data that will settle" arguments between dischargers and environmentalists. Bascom did just the opposite.

Isaacs was probably thinking of Rim Fay when he mentioned opposing environmentalists. Ironically, during Bascom's travels, he met a marine biologist named Ed Ricketts, famous for John Steinbeck's portrayal of him as Doc in *Cannery Row*. Bascom admired him: "Ed was a kindly, easygoing fellow," he wrote in his autobiography, "who cared little about money but a great deal about the finer things in life, including women, beer, Gregorian chants, and sea animals." Fay saw himself as another Ricketts (although not so enamored of Gregorian chants), committed to the environment, marine critters, hard drinking, and the occasional seduction.

Just the same, Fay and Bascom disliked each other, verbally duking it out for years. "Both [Rim Fay] and Bascom came out of the same mold," Bruce Thompson says. "They were both full of bravado and they both thought they knew much more than they do. So I can see why they didn't get along." Fay and his close friend Don May followed Bascom to each public hearing at which he spoke, sometimes carrying their sad samples of tumor-covered white croakers and disputing nearly everything the man said. Fay was the go-to guy for reporters who needed an opposing quote if Bascom said the bay was in fine shape. And if Fay told journalists that sewage and DDT had killed the bay, they called Bascom for his dependably contrary quote.

This occasionally shrill debate somehow never penetrated Los Angeles' City Hall in any meaningful way. Bascom's summaries to SCCWRP's reports diluted the reports' own contradictory data with slithery language that hinted that the sewage discharges weren't necessarily harming sea life to the extent people such as Fay said they were. Bascom's point of view, formulated to secure the 301(h) waiver, trumped all others. Mayor Tom Bradley's administration, the city council, and the Bureau of Sanitation's engineers were all swayed by Bascom, the charming man with the resonating voice and authoritative silver hair.

And there he was, the villain in Bennett's developing morality play. But for the moment, the schoolteacher didn't know Bascom existed.

CHAPTER 5

The Press Conference

Howard Bennett was a walking contradiction. With a sometimes lofty Ivy League lilt, he would quote Shakespeare to buttress a point he was making, and yet he saw himself as the common man, and one who understood what other guys supposedly just like him wanted. While he'd never cozy up in a lounge chair with a beer, he believed he knew what average Joes would respond to. He had a special talent, he said, for the kind of publicity that appealed to the masses. The same masses who probably never read Shakespeare or had heard the quote "Something is rotten in the state of Denmark."

However, the only efficient way for Bennett to bring his message to the ordinary folk, who would no doubt recoil from the idea that they took their kids to a polluted ocean, was through the press. For guys like Bennett, the media had but one principle: a little sensationalism is a sweet thing. Give them a crisp, clear creed spiced with threats of personal doom and you can hold an audience's attention far longer than the city council can while debating building codes. Throw in the *ick* factor that any-thing having to do with sewage entails, and you've got every television journalist's dream.

And what about readers perusing the newspaper? Bennett wasn't so concerned with the print media. He understood how powerful images and sound can be when you want to wake up a sleepy metropolis. He also figured more and more people got their current events from television, and so he put most of his effort into soliciting the unwashed masses who sat bug-eyed before their Sonys.

Of course, before he could do that, he had to reach the news gatekeepers—not always the easiest task. Bennett told people he had developed a sense of what attracts reporters when he briefly worked as a copy boy at the *Wall Street Journal* in the early 1950s. He painted the image of a naive kid who got the job through a girlfriend or a girl who was a friend—he wasn't too specific—and indicated he had been schooled by hard-core, jaded journalists who wouldn't bother with a story unless it grabbed them by the throat. Just to make the narrative juicier, he added that he had written a play review when the theater critic couldn't make the show and gave Bennett his ticket. It was a great tale, but he later conceded that he probably didn't learn that much from the staff there after all.

Perhaps closer to the truth was the lesson he intuited after momentary fame in 1951, when he hitchhiked twelve thousand miles from New York City to Anchorage, Alaska, and then back, all in seventeen days. After taking on a two-hundred-dollar bet he couldn't do it, the twenty-one-year-old Bennett started on August 1 at the Holland Tunnel, made his way to Los Angeles to visit his parents, and then headed up to Anchorage, where he sent out postcards to friends declaring, with the kind of hyperbole he would later perfect, "What have I done? Why have I come here? This is indeed the edge of the world!"

He even worked the visual angle by going to the 3M Company before the trip and talking them into making him a two-foot-by-four-foot sign with ALASKA on one side and NEW YORK on the other, using a new reflective material they had developed for road signs. He promised them he would talk up the product should the subject ever arise. The sign not only got him a lot of rides but also turned into a great prop for newspaper photos.

In an era when Alaska could have been the moon, Bennett turned the experience into a publicity bonanza. Newspapers in New York and in the cities he passed through featured his exploits. *Parade* magazine devoted two pages to his adventures. The national CBS radio show *It's News to Me* interviewed him (they paid him twenty dollars for his time), and it was there that he truly exercised his developing talent for telling stories.

A year later, he collected hundred-dollar "investments" from friends and thumbed his way through South America, establishing at the time a record for hitchhiking some thirty thousand miles. His stories were broadcast on the Voice of America, and he appeared in *Parade* for the second time. Here was a guy who figured out early how to get press coverage by putting himself in reporters' shoes and telling a story they would like to retell. He also learned just how far a good story could go; *Parade* in 1952 claimed a readership of more than 13 million people.

And so, with another story to tell, Bennett, using his late father's Underwood, slowly typed a manifesto of sorts concerning

Santa Monica Bay pollution. The resulting three-page document was sometimes more screeching indignation than reasoned argument, with several skewed facts to help turn up the volume. He announced that "L.A. wants to dump more untreated sewage into the ocean," and "L.A. wants to save money on the lives of our children." Ahem. Actually, as mentioned before, Hyperion was dumping partially treated sewage into the bay five miles offshore, which was indeed destroying the marine environment but, according to marine biologists both then and now, had little if any impact on human health (well, so long as no one ate the fish, but that was another story; Bennett skirted the DDT and PCB issue altogether, as it was somewhat irrelevant to the waiver).

The declaration went on to one of Bennett's major peeves, the apparent lack of adequate notice for the March 25 hearing. "The general public," he wrote, "was cheated by making it almost impossible for the public to attend the critical Joint hearing on March 25. It was deprived of its right to know! The hearing was sparsely attended—only 5 people showed up to testify. It wasn't lack of interest. It was lack of notification. VIRTUALLY NO ONE KNEW ABOUT IT! Those that did have their names on a list were given 48 hours' notification. Even a condemned murderer is given more time to prepare his appeal."

With a natural ability to compose provocative statements, Bennett concluded, "We feel the entire problem literally stinks to high heaven. The Santa Monica Bay is being used as the toilet bowl for millions of people. It has some of the worst pollution in the world. Why must our children be forced to swim and play in it this summer?

"Mayor Bradley—Why don't you try to grow your goldfish in a toilet bowl and see what happens!!"

While this wasn't necessarily the most logical thing to say, it did point to the only person Bennett knew at the moment to blame for the pollution. He instinctively realized that he could get only so much mileage from attacking the faceless Regional Water Quality Control Board, EPA, or city. He needed a single, easily visible target. And that was the mayor.

Bennett's anger had now been turned into a series of catchy phrases suitable for quoting, and so it was time to share his message. He figured this would require television cameras, and the fastest way to bring them before his snarling face was to announce a press conference. Following a certain logic—that he should go where the media lives—Bennett booked a room at the Los Angeles Press Club on Vermont Street. There, for several hundred dollars—of his own money—he rented a large enough space for reporters and cameras, plus a breakfast buffet. He didn't know it at the time, but even failing a good story, some reporters could still be lured by a little free food. (The Press Club also had a bar for those reporters supposedly off the clock.)

On April 4 at ten o'clock, Bennett waited for the press to arrive. He had asked Ruth Lansford to join him, as well as an attorney, Barbara Blinderman, who provided her time pro bono (Bennett had been advised by someone apparently unfamiliar with press conferences to have a lawyer present for legal backup). The three sat at a long table on a small stage with three easels behind them holding poster-sized pictures of trash on the beach. This didn't really have much to do with sewage, but it helped add to the bleak message Bennett was about to introduce.

At this point, Bennett's coalition claimed ten member organizations, most of which would be expected to protest ocean pollution, including the Friends of Ballona Wetlands, Catalina Swim Federation, and the Playa del Rey Homeowners Association. But down near the bottom of the list was the Southern California Jujitsu Association, included because Bennett needed as many names as he could get. Bennett was the group's registrar, and his son, Leif, was also involved in it, and the two of them convinced the president that it made sense to put the association on the coalition's letterhead.

Don May's group, Friends of the Earth, had been on the first letterhead, but had been taken off a week later when they splintered over the firing of their leader, David Brower (who had helped found the group in 1969 and, before that, had made his enviro reputation as the Sierra Club's first executive director). Also missing was the local Sierra Club chapter, which refused to be included; however, the chapter did organize its own Clean Coastal Waters Task Force about three months later under a feisty woman named Nancy Taylor, a close friend of Rim Fay's.

Two television stations, KABC and KCBS, showed up, along with two news syndicates, City News Service and Copley News Service, plus a news radio station, KNX. And that was about it.

The City News Service had sent a young reporter, Marc Haefele, to cover the press conference. Haefele saw on the Press Club's stage "this very gangly type with an incredibly earnest demeanor, with the schoolteacher's didacticism and a really odd accent I have yet to figure out." With city affairs as his beat, Haefele would follow Bennett's activities from this point on.

Haefele must have also noticed Bennett's eyebrows, two curly explosions of hair like little hedges below his balding head. Barbers wanted desperately to trim them and he refused, as though he knew his eyebrows helped rivet one's attention to him when he told his stories.

Wearing a blue windbreaker over a yellow T-shirt, Bennett began his remarks, reading off his manifesto. "Let's talk about the problem," he said earnestly. "The fish are dying and the sand crabs have almost disappeared. Because of ocean sewage dumping, most of the Santa Monica kelp beds are gone. The perch didn't spawn this year. The Pismo clams that used to be pearly white—" he paused for effect while his eyes scanned the room, and then snapped, "have black shells now. And the lifeguards have gotten cancer from dumped toxics." Spiking his delivery with just enough righteous indignation to let them know he was sincere, and yet to not seem shrill, he added, "The ocean is now a critical mess filled with a stinking mass!"

In his everyman clothes, with his thinning hair combed back and thick, owl-eyed glasses, he looked and sounded a little like a well-educated fisherman. As he slightly hunched over his notes, however, the schoolteacher in him seeped out, castigating Mayor Bradley as if he were reprimanding a failing student. His voice dripped with authority and scorn, and he raised it a little louder each time he made a point, as though each succeeding idea made him angrier. However, he resisted breaking out the Shakespeare.

Given Bennett's accusations, both television stations contacted Harry Sizemore, the Los Angeles Bureau of Sanitation's assistant director, for his comments. Surrounded by gurgling sewage at the Hyperion Sewage Treatment Plant in the KCBS-TV report, Sizemore put in the city's pitch for saving money, similar to what he told the water board on March 25. "If we went to full secondary," he said, "we would have to spend another $150 million for capital construction, and our operational costs would go up another $15 million a year." Sizemore, with his carefully combed blond hair and round face, appeared surprised anyone would bring up the subject.

However, before KABC-TV's cameras, a defensive disdain seeped through. "We are abiding by the existing permits for operating this treatment plant, but we have a very active program for improving our processes," he said. "We're currently spending $200 million upgrading our solids handling, which will have a dramatic improvement in the amount of solids—particularly heavy metals, toxic organics—going out to the ocean. We expect that to be operational by the spring of 1986."

Importantly, in the public relations message that Sizemore tried to cheerfully present, he obliquely admitted that the city had been sending into the bay effluent that was better left out of the already polluted water. If nothing else, Bennett had accomplished one important objective—he had made someone at city hall squirm a bit.

Just the same, Bennett was more interested in a different goal, as he told the press conference: "We want another, publicized public hearing, a joint EPA–Regional Water Quality Control Board. We want one more chance." He then added with a tone so rational one would think this was the most reasonable request, "Just one more chance to have the public testify on this particular issue."

Given that his campaign had started only a week before, it's perhaps surprising that Bennett's plea for a second hearing was already being considered by—among others—the EPA's chief of the oceans and estuaries section, Patricia Eklund, whom Bennett had called with his complaints. A decision would come a week later.

CHAPTER 6

City Hall

In an open letter sent out to coalition members and other interested parties on April 11, Howard Bennett asked Los Angeles mayor Tom Bradley, regarding his administration's consistent assertion that full secondary treatment at Hyperion would cost too much, "Is saving money more important than saving the lives and health of people?"

According to those who worked for Bradley, it wasn't that he wanted to see Santa Monica Bay polluted any further in exchange for saving a few million dollars. Instead, the guy was running a huge city, and he had staffers to handle such matters. Insulated from Bennett's accusatory tone, Bradley considered the 301(h) waiver the kind of esoteric administrative detail that should cross someone else's desk, not his.

As Mike Gage, Bradley's deputy mayor from 1987 to 1990, puts it now, "I think it's fair to say that the mayor paid attention to the items that were either brought to him by his staff, by things that he interacted with the community on, or by reading things in the newspaper. He used to ride his [stationary] bike every morning about 5:30 to 6:00 A.M. and read the *LA Times*. I asked him once when things were a little bit rough how it went. He said, 'Every time I [read something I don't like], I just pedal faster.' So he got his information from a lot of sources. And certainly, I think it was incumbent upon his staff, of which he had a hundred-plus, as well as the city employees, which were 30,000-plus, to bring things to his attention in a broader fashion than he would be able to get his arms around just as a single entity."

In other words, you might be the most powerful man in one of the nation's most powerful cities, but you can't know everything that happens around you, especially if it's some piddly waiver that—at the time, at least—had no political weight. Instead, you have people to do the knowing about such things for you. But it would seem that, in the case of Hyperion, the people who did the knowing didn't know enough. Like potholes on a back street, the 301(h) waiver issue didn't cross staffers' desks beyond a certain power level. This was something that specific, relevant departments addressed, and no one thought it needed to float up to the mayor's sphere. Still, one wonders how Bradley managed to avoid the subject.

By 1985, Bradley had been the city's mayor for twelve years, but it's safe to say that, long before 1973, when he first won the job from Sam Yorty, the environment had rarely crossed his radar. Born a sharecropper's son in Calvert, Texas, and the grandson of former slaves, Bradley came to Los Angeles when he was seven. At age twenty-three, after attending high school and some college, he joined the Los Angeles Police Department in 1940. After earning a law degree from night school, he was elected to the city council in 1963, from which he launched his first mayoral campaign in 1969, running against Yorty, who infamously suggested Bradley would invite militant black nationalists into town if elected. The same white voters who had supported Bradley in the primary fled to Yorty and reelected him mayor. Four years later, Bradley tried again and this time won.

During that time, the man who managed to ride his bike and read the newspaper at the same time may have skipped over the occasional article covering the bay's pollution. In 1974, the fifty-seven-year-old mayor might have seen a *Los Angeles Times* piece on the "sea of sludge" in Santa Monica Bay and the environmental havoc some said was the result. However, in this article Robert Bargeman, then the Bureau of Sanitation's director, stepped lightly around the notion that sludge could create a dead zone. "There are not as many species of sea life on the bottom," he conceded. "It's not a dead area—there's just not the variety there was. But it's a big ocean out there, and we're talking about a one-square-mile area."

This sounded suspiciously like something SCCWRP's Willard Bascom could have said (he'd begun work at the four-year-old organization a year earlier), and if Bradley had seen the news report, he probably would have been satisfied that his staff didn't think there was a real problem. It would seem that city hall was an insular place where Bradley put much more stock in the staffers hired to know a topic than in those from the outside who were also hired to know about the same topic—and who occasionally disagreed with his people. According to a staffer who worked there in 1985, "Bradley's method of governing was very much different, I think, from other mayors, in that he would assign responsibility to people, listen very carefully to

Los Angeles mayor Tom Bradley after his election on November 4, 1986.

what they had to tell him, and then he would act on their advice, but not necessarily advise them of this decision prior to making it public."

It was *someone's* decision to continue the sludge discharge. However, two years later, the State Water Resources Control Board sided with the antisludge contingent and required the city to stop dumping sludge by April 1, 1978, as part of its NPDES permit that allowed it to discharge treated sewage into the ocean. The city council approved a ninety-six-million-dollar bond measure to pay for the facilities needed to handle the sludge, which would be funded by increasing sewer service charges. Voters defeated the measure, partly because of their resistance to higher taxes and partly because the city council itself openly resisted the idea that sludge should go anywhere but in the bay. For the record, Bradley supported the measure and, at the time, apparently didn't worry about upsetting voters with higher sewer bills; perhaps he was lulled by past elections in which they had passed similar bond measures. Without apparent fear of the contradiction, he also supported his staff's efforts to fight the state's and EPA's sludge-out requirements.

At the time, the city council had based its opposition on a 1973 SCCWRP report (its summary written by Bascom), which concluded after three years of research that there was no need for any "substantial modification of current wastewater disposal practices." In other words, dump away—it wasn't hurting the marine life. And, oh, by the way, dilution is the solution to pollution.

This led Councilman Zev Yaroslavsky in 1976 to write an editorial in the *Times* blasting the idea that the city needed to take on the expense and headaches of disposing of sludge anywhere but in the ocean. After quoting the "best scientific opinion," he painted a bleak picture of hundreds of trucks transporting sludge to landfills as though it were plutonium. "The liquid, bacteria and toxic waste content of the sludge will render the landfills useless for decades to come," he warned.

Yaroslavsky says now that his opposition was partly based on the federal government trying to convince the city to install an untested process for drying and incinerating sludge. "I thought we were buying a pig in a poke," he says, and, in the article, called it the "EPA's mad scheme" for proposing that the city, already choked with air pollution, add more smoke to the skies.

He says he was also relying on a Los Angeles Bureau of Sanitation official, William Garber, and SCCWRP's Willard Bascom, who both told the city council that the sludge wasn't harming the environment. On November 13, 1981, the Bureau of Sanitation even took him and other council members out on their research boat, the *Marine Surveyor*, to prove their point. It's hard to imagine a dumber demonstration of how wrong the city's position was. Near either the five-mile or seven-mile outfall (Yaroslavsky doesn't remember which), the boat's crew dragged a net through the water for fish and managed to bring up a horror show of fin rot and tumors that they dumped on the boat's deck. "They didn't look healthy," Yaroslavsky says. "They didn't look like the fish that were in my goldfish bowl. [And that caught my attention.] The most dangerous thing you can do to a public official is take [him] out to prove a point, only to have that point refuted in front of [his] eyes. That's what happened that day. I will tell you that, as far as my buying Garber's and SCCWRP's line . . . that was a public policy mistake that I regret to this day."

If the debate over sludge didn't raise the issue's profile enough for Bradley staffers to bring him up to speed on the topic, an EPA lawsuit in 1977 claiming the city had violated the terms of its 1975 NPDES permit should have. The suit charged that the city had consistently dragged its heels on dealing with the sludge issue, missing deadlines for, among other things, submitting a plan to stop discharging the "liquid, bacteria and toxic waste" (as Yaroslavsky put it) into Santa Monica Bay. That deadline, by the way, had come from another lawsuit three years earlier, which forced the city to study how best to dispose of the sludge. Granted, people sue cities all the time for any number of reasons, and Bradley's legal staff no doubt didn't bother him with all the details. However, when the fed's lawyers come knocking, not once but many times, the mayor should be aware of it. And so the question lingers—just how interested was he in the topic?

Or, as Gage puts it, "I think [Mayor Bradley] was always open to and relatively sensitive to environmental concerns across the board. I think that if you're running a city the size of LA, there's just a whole lot of pressing priorities, and some get your attention and some don't. It took a while."

Los Angeles pretty much stood alone among coastal cities in its resistance to the EPA's sludge-out requirements. However, forced by the 1974 lawsuit, the city, along with Los Angeles and Orange Counties, jointly began to study the issue. Three years later, just before the 1977 EPA lawsuit was filed, the committee recommended an old idea—burning the sludge—along with a state-of-the-art way to do it and generate electricity at the same time. They called it the Hyperion Energy Recovery System, or HERS. (This was the program Yaroslavsky opposed.)

At the time, HERS probably sounded brilliant. Sludge, which contains a great deal of water, would first get a spin-dry treatment in centrifuges and then, to wring out every last drop of liquid, go through a patented system called the Carver-Greenfield Process. The resulting powder, when burned, would fuel huge steam turbines to create electricity to power the plant. Just the same, there was some hidden chutzpah in all this. It seems this was slightly experimental, in that the Carver-Greenfield Process had never been applied to sewage sludge. Rather, it had been invented in the late 1950s by Dehydro-Tech Corporation for animal rendering and was later applied to wastes at a brewery.

Still, this wasn't just a technological answer, but a financial one as well. The EPA, per the Clean Water Act, which included funding incentives, would pay the bulk of the system's $185 million cost, with Los Angeles picking up just $4.625 million. Moreover, this would please the EPA, which wanted to implement the Carver-Greenfield Process *somewhere*.

Even with the city—and EPA—touting HERS as the breakthrough sludge removal system, it still took three years for the Bradley administration to cut a deal with the feds regarding the sludge lawsuit. In the consent decree, signed in June 1980, Los Angeles agreed to stop dumping sludge in the ocean by July 1, 1985, and, at the same time, to develop a construction schedule for HERS. The EPA fined the city $2.6 million for previous violations of its NPDES permit.

In September 1979, the city had also sent in its application for a 301(h) waiver (endorsed by Willard Bascom). Not surprisingly, the document, prepared by the Bureau of Engineering, put a positive spin on the city's ability to comply with the waiver's eight basic clean water requirements. The EPA hired an environmental consulting firm based in Bellevue, Washington, called Tetra Tech, to evaluate the application for its veracity. Tetra Tech wasn't impressed.

Bulldog that he was, Howard Bennett sniffed out the 1981 Tetra Tech evaluation and gave it to his son, Leif, to read through, as though the technical jargon the report contained would flummox a simple schoolteacher. "Hey, he graduated from Caltech," he told people. "And they don't take dummies." Leif had a bachelor of physics degree, and even though he had considered bumping up his education and getting a master's, he instead had taken a job in Xerox's computer division. Already a proud father, Bennett was truly impressed, figuring that anything involving computers was so cerebral as to be beyond his limited intellect. In fact, he consistently downplayed his own cognitive prowess, as though he had faked his way through life and was just waiting for people to finally believe him when he claimed he wasn't that smart and had never been an exemplary student.

So Bennett, thinking he had the smoking gun, brought out his smarty-pants son to analyze the evaluation. "The report was very nicely written from a publication standpoint," Leif Bennett says now. "Not exactly a nonscientific publication, but accessible if you didn't mind science. . . . Very straightforward, easy to read. It was really easy to find where [the application's data] was outside the standard, and relatively easy to find ones that were far outside the standard."

This included one of the most crucial items, the 301(h) waiver requirement that a balanced indigenous population, or BIP, must exist beyond the zone of initial dilution, or ZID. In essence, this meant that, beyond the area where the treated sewage spread out on the ocean floor—the ZID—species diversity should remain stable and, in the best of all worlds, the same as it was before the discharge began. However, the Tetra Tech report concluded that the city's planned improvements to Hyperion wouldn't be enough without also going to full secondary treatment. "Non-BIP conditions are predicted to extend far beyond the ZID if all proposed improvements, including termination of the sludge discharge, are implemented."

Nevertheless, the city's application even gave BIP status to the ZID: "The discharge area supports a fish fauna that would not be markedly different in the absence of the discharge," it said. This was already known to be untrue, in that the species diversity had declined to but a few species, all of which were adapted to the organic-rich environment. The Tetra Tech people were apparently loath to actually say this was a deliberate fib, but gently wrote, "The available information indicates that local differences in demersal fish and epibenthic macroinvertebrate communities [the small creatures that live in the muck, known as the benthos] are associated with the effluent discharge."

As Leif dug further, he found instances where the application provided data that demonstrated Hyperion wouldn't come close to meeting EPA standards, and one has to wonder if the city's authors were just trying to slip it past the EPA and its reviewers. For example, the EPA used biochemical oxygen demand, or BOD, as a measurement of the organic material coming from sewage discharges, which sucks up oxygen wherever it's deposited. The higher the BOD, the less oxygen there is for organisms in the discharge area. The report said, "The BOD in the effluent of the Hyperion treatment plant's five-mile outfall exceeds the secondary treatment criteria described in

40 CFR 133.102(a), which specify a BOD removal efficiency of at least 85 percent and 7-day and 30-day maximum effluent BOD$_5$ levels of 45 mg/l and 30 mg/l, respectively. . . . For the improved treatment plant, the designed BOD removal efficiency will be 61 percent, as shown on the NPDES permit application, and the annual average effluent BOD$_5$ level is predicted by the applicant to be 105 mg/l."

Finally, the report brought up the issue of monitoring. At the 301(h) waiver hearing in 1985, Rim Fay had predicted the city couldn't be trusted to do its own monitoring, and the report essentially said the same thing: "These deficiencies in sampling design and the absence of stressed control stations as part of the self-monitoring program severely limit the utility of self-monitoring program data in conducting BIP comparisons and predictions." In other words, Leif concluded, the Hyperion folks had designed their self-monitoring process so that it wouldn't detect "the worst effects of various forms of pollution," as he wrote in his analysis of the analysis.

The Tetra Tech report, in its more than four hundred pages, yanked out example after example of how the city's application fell short of 301(h) waiver water-quality requirements, and, perhaps worse, how it sidestepped the darker data to make the environmental situation appear better than it was.

Despite Tetra Tech's grim analysis, the EPA tentatively approved the application.

About two years before Tetra Tech's report, Bradley appointed his campaign fund-raiser, Maureen Kindel, as president of the Board of Public Works. This meant she oversaw Hyperion just as the sludge issue was shifting into high gear, and it put her in the position of disagreeing with her sanitation colleagues who believed Hyperion wasn't destroying the bay's ecosystem. "I felt there was just no way to justify the pollution in the bay and our contribution to it," she says. "It didn't seem to take a rocket scientist to figure that out."

But, in spite of the Clean Water Act's requirements, the engineers who worked for her had fixated on the mantra "Dilution is the solution to pollution," and according to her, it took time to convince them otherwise. Kindel says she found them to be largely a tight-knit group of rigid, arrogant, overbearing men who rejected any notion that their beloved, though broken-down, treatment plant was doing anything to harm the environment or human health. It was like accusing them of not doing their jobs. And, being a woman in a waning era in which men could still get away with snubbing a female colleague, even if she was their boss, Kindel assumed her sex worked against her ability to convince them that primary treatment wasn't doing a whole lot of good.

"As president of the Board of Public Works," she says, "with no particular training in any of this, I had to take my time to find out what was going on. And to be honest with you, it probably took about two years or so. First of all, I was the first woman ever to serve in that position in the history of the city. And it was a male bastion, and all the people that worked for me were men. Most of them had joined the city after the Second World War. All of them had engineering degrees, which I did not have. And I had to figure out initially what my role was in this."

In particular, she couldn't get through to Don Tillman, chief deputy city engineer, who was at that time the country's undisputed godfather of sanitation engineering. Kindel was in charge, sort of, but by her reckoning Tillman didn't really seem to care. "I remember that Tillman would always say, 'Dilution is the solution,'" Kindel says. "And [he would say that] this was all the imagination of these environmentalists—this was always said in a derogatory way. Because from [the engineers'] perspective, the environmentalists . . . were not technically trained, and, you know, [the technical people] were the ones with the engineering degrees, and they were the ones that knew the science of it all. So time marches on, and the situation doesn't get any better. And we have the city attorney's office in here fighting the regulations, as well as the technical people saying that this isn't a problem."

Kindel says she could have talked to Bradley about it—and did—but typically, the engineers got to him first, telling the mayor to fight the EPA. "Bradley was a conservative guy from that perspective," she says. "He was getting advice from his so-called professionals. He would pretty much follow it, unless he had a good reason not to follow it."

Ultimately, however, the argument—the public one, at least—came down to money. In a *Los Angeles Times* article, Bradley's administration and the city council, as if making a threat to voters, warned the public their sewer fees would jump from an average $5.40 a month to $10.60 to pay for any new construction required for full secondary treatment. As Joy Picus, a city councilwoman, put it in the same article, "We would be looking at doubling the charge . . . for a negligible benefit." Or as John Dorsey puts it now, "It's very expensive. You don't make a profit in this industry."

By 1985, when Bennett turned up the volume on this debate, Los Angeles ballyhooed the $180 million it was already spending on Hyperion improvements. City officials complained that full secondary treatment would cost an additional $155 million, and that it wouldn't substantially improve water quality. And finally, there was HERS, which at this point the city now projected to cost $200 million while seldom mentioning that most of the money would come from federal funding.

The city waved these dollar signs whenever possible to justify its position. Meanwhile, Mayor Bradley said little himself, depending on his staff to defend the city's position. After all, that was their job.

CHAPTER 7

The Activist

As a storyteller, Howard Bennett isn't one for analyzing pivotal moments in a narrative—those plot points where, if a certain thing didn't happen, the world would be different. He simply cruises through the story, incorporating as many asides as he can squeeze in to explain one thing or another, until the tale starts losing steam. If a certain event in the story has a meaning, he chooses not to think about it. The story itself is enough for him; he doesn't need to get all intellectual about it.

That's one reason why—when talking about the coalition's start—he tends to race past his first encounter with Dorothy Green, then-president of the League of Conservation Voters, Los Angeles chapter. The nine-year-old league was a political clearinghouse of sorts that endorsed politicians based on their pro-environmental stands, and although the group didn't get involved with individual issues, they were still an obvious contact for Bennett to make. Using the phone list Ruth Lansford had given him, he dialed Green with no idea who she was, what the league did, or how that call would later affect so many people. And when he tells the story today, he still doesn't mention that last item, however dramatic it proved to be.

Like most environmentalists at the time, Green hadn't heard of the 301(h) waiver. It's fair to say that she didn't make a connection between flushing a toilet and Santa Monica Bay pollution. While it was now common knowledge that DDT and PCBs covered parts of the seafloor, nobody at the time knew what to do about it, and so it was hardly a topic to rally public outrage. As for sewage, if anyone even talked about it, the Hyperion plant's deficiencies weren't widely known, nor were they something you could easily stick on a picket sign. When Bennett called Green, he not only informed her of a situation she hadn't realize existed, he also gave her a pithy handle on the issue—oppose the waiver, stop the pollution. As she put it, "Howard notified us. He notified the press, he notified academia, he notified everybody—Howard Bennett was a real publicity hound."

For the moment, the pure shock over the waiver was enough to motivate Green. After she put the issue before the league's board and they agreed to join Bennett's coalition, Green sent out a letter, on April 11, addressed to the chairs of the state Environmental Quality Commission and the Board of Public Works. While not as strident as Howard Bennett's missives, it nevertheless carried a stern, impassioned tone that reflected her surprise that the Clean Water Act had a deliberate loophole:

"We are asking that both the Environmental Quality Commission and the Board of Public Works immediately start investigations of the waiver process. The residents of this city demand to know why the waiver is being pursued. We demand to know

Heal the Bay's Dorothy Green in 2008. She helped found the organization, using Howard Bennett's coalition as a start.

who is responsible for this policy that would subvert the Federal Clean Water Act and that ignores an existing court order to clean up our mess. We also demand to know why the information given to the EPA in support of this waiver application is so deficient."

While it might have been a little presumptuous for the tiny league to speak for the rest of the city, the letter in its modest way had the right amount of indignation for an opening shot. But Green's crew would have little effect with just a declaration of anger, so the league's board gathered at her house, a comfortable circa-1928 Tudor near the UCLA campus, where Bennett joined the group for strategy sessions concerning what to do next.

With the eyes of a poorly paid schoolteacher, Bennett entered Green's house for the first time as if it were the Emerald City. Her nearly four-thousand-square-foot residence dwarfed his boxy beach home. Awed by the heavy wood front door, he walked through the hallway with its fabric-covered walls and was engulfed in a warm atmosphere of wealth. Green and her husband, Jack, who made his fortune in construction, bought the house in 1966 as a fixer-upper and remodeled it, and even though neither liked the place's Old World feel, they never moved.

The board usually met in Green's living room, a long, high-ceilinged space with exposed, rough-hewn wood beams and white plastered walls. One half expected to see the English moors through its windows. Here, sitting on two couches plus chairs taken from the dining room at the far end, board members listened to Bennett outline what he knew so far and his ideas for getting the issue out.

As far as he was concerned, his initial press conference constituted only a mild first crack at publicity. Bennett wanted to keep hammering at the issue. He believed confrontational rhetoric that incorporated a visual feature to attract television news cameras always worked best. With an angry zealot's enthusiasm, he threw out idea after idea to the astonished group, outlining a stream of demonstrations that the city, EPA, and Regional Water Quality Control Board couldn't help but notice, or so he thought.

Among those at the meetings was the league's forty-five-year-old vice president, Moe Stavnezer, a gregarious pharmacist by day and gadabout activist in off-hours. (Don May and Martin Byhower—who were involved in other groups as well—also attended some of the meetings.) According to Stavnezer, Bennett suggested they storm Los Angeles International Airport and greet airline passengers with information about the waiver like environmental Hare Krishnas. The group objected to that idea nearly as soon as it came out of Bennett's quick-talking mouth.

Stavnezer believed Bennett saw them not as collaborators but as facilitators for his plans. The schoolteacher was intent on doing things his way, and it appeared he wasn't interested in dissenting views. "Howard had his own ideas about how things should go, and he pushed those things very hard," Stavnezer says now. "I don't remember whether or not they were rejected at those meetings, but I do know the folks from the league and some other people thought Howard was way off base about a number of his suggestions. And that it became clear that there was not going to be a coalition through Howard Bennett."

Green sat on an antique Louis XV bergère armchair, its back against a window, listening to Bennett toss out his schemes like hand grenades. At fifty-six years of age, she was a slender, prim woman, neatly dressed, with her hair pulled back. She didn't object to a demonstration or two, but in principle preferred a more measured, mannered approach. Whereas Bennett saw nothing wrong with thrusting a middle finger toward those in power, she developed relationships even with the opposition, figuring that a seat at their table meant more in the long run than a quick stab in the eye via a photo op. Sure, she admitted, you needed the press coverage—not just television but radio and print as well—but if all you did was yell insults, those in charge would shut the door. It had happened to Rim Fay, who was usually more bellicose than the decision makers liked.

Green had developed this attitude over the previous fifteen years. In 1970, she had been raising three boys, happily married to Jack, and yet rather depressed. If asked about it at the time, she would take great pains to explain that her personal life had nothing to do with the melancholy. It was more that she looked around her and saw that her son had just registered for the military draft; the civil rights movement seemed to be ending, for better or worse; and the first Earth Day, in 1970, reminded her how banged-up the environment had become. She had become so dispirited that she crawled into bed—both figuratively and literally—to escape it.

Fortunately, this didn't last long. Green quickly realized that, as a long-term strategy, self-pity was limiting and a better response involved action. Not one who believed success came out of multitasking, she decided to tackle one cause at a time, and, influenced by her cousin Steve Beckwitt, a young Berkeley enviro, she chose the environment. From there, she shopped for an organization to join and found Women For:, a Beverly

Hills–based group of women who wanted to be politically involved beyond the traditional coffee-making and stamp-licking
duties they had been given thus far in election campaigns.
The organization's name, while looking like a typo, was deliberately open-ended. "They were for education and peace and
the environment and all the good things in life," Green said.
"And so I found myself heading up the environmental action
committee."

This led first to working on the successful passing in 1972 of
Proposition 20, which established a state Coastal Commission
to oversee coastal development. She also lobbied in Sacramento
for environmental legislation, where she learned firsthand the
role money plays in politics. While she and her two Women
For: mentors were eating dinner at a legislator hang-out, a Chi-
nese restaurant called Frank Fat's, the speaker of the assembly,
Bob Moretti, spotted them and picked up the tab. For the record, the legislative bill that the group supported would have
created an agency for setting environmental policy for the rest
of state government, and it died as a result of what Green believed were oil interests doing their own lobbying.

Green's activism over the years took her to water use issues,
always a touchy and complicated subject in California, given the
tug-of-war over limited water supplies for agriculture, water for
cities, and water being stolen from other regions. (Los Angeles
infamously hijacked water rights from the Owens Valley, north-
east of the city, and, at the time, was still enjoying the flow that
resulted.) Water use issues consumed her like no other topic,
and even though it was a stretch to connect the 301(h) waiver to
her passion, she took it on. If nothing else, the waiver had an
easily grasped quality to it that other water issues didn't.

Yet even though one early coalition press release ended with both Bennett and Green listed as contacts, the two maintained a frosty affiliation. Green, with her instinct to build relationships, tolerated Bennett's belligerence, while Bennett didn't pretend to abide any of Green's thus far gentle suggestions that he tone it down. In a sense, they needed each other for a brief time. A very brief time.

CHAPTER 8

The Second Hearing

Considering what Rim Fay had said about the first 301(h) waiver hearing—that there was no way Howard Bennett could reverse the waiver's inevitable approval—getting a second hearing before the EPA and Regional Water Quality Control Board was surprisingly easy.

Less than two weeks after he announced, on behalf of his nascent coalition, "We want one more hearing—just one more chance," the EPA, Region 9, sent Bennett a letter dated April 9 that, in the most neutral tone, complied with his somewhat pugnacious request. The one-page missive hardly took a whole breath to read out loud: "As per your request, a second hearing on the draft section 301(h) modified NPDES permit for the City of Los Angeles Hyperion Treatment Plant will be held on May 13, 1985, at 10 A.M." The rest of the brief letter outlined the meeting details, and it was signed by Patricia Eklund, chief of the EPA's Region 9 Oceans and Estuaries Section. Thirty-three years old, Eklund had just taken the job after working for the

EPA since 1977 and, before that, the Army Corps of Engineers. She says now that she and the Regional Water Quality Control Board's executive officer, Robert Ghirelli, jointly decided to hold another hearing based on public outcry.

However, if Bennett had expected a conciliation along the lines of "Gosh, your point was well taken, and, yes, you deserve a second shot," there was nothing in Eklund's letter that suggested it. Despite the stoic tone, Bennett sent Eklund a red rose bouquet as though they were secret compatriots. Eklund says she never got the roses, but, per federal rules, she would have had to refuse them anyway.

(Bennett has always maintained—without substantiation—that the EPA had never before reopened the consideration of a 301[h] waiver to further public comment by holding a second hearing. Eklund, however, isn't so sure the meeting was as unprecedented as Bennett wants people to believe.)

Although he sent the celebratory roses, Bennett barely smiled at the announcement. A second hearing meant nothing unless he stacked the deck so fat with aces that the city couldn't win. That required not only filling every seat in the hearing chambers but also, if he could, bringing a vociferous throng so large that the bodies spilled out into the halls. He wanted such thunderous indignation hurled in the water board's and EPA's direction that they wouldn't dare approve the waiver application for fear of a citizen revolt.

But this couldn't be just any angry mob hissing from the uncomfortable chairs. First, it had to be people who could give cogent testimony and pound the facts over and over—Santa Monica Bay is polluted, it's a menace to marine and human health, and how dare you even suggest that the city's idea of adequate sewage treatment is the primary treatment once-over. But, second, there had to be an unspoken message in there, something beyond rational argument that would influence the decision makers. He needed those aces.

In what's been considered his most brilliant strategy, Bennett tapped the one natural resource Culver City High School— where he worked—offered: the students. If nothing else, most of the kids regularly visited the nearby beach, and many were surfers who blanched at the idea of sliding under a wave's curl loaded with what Bennett had them believe was pure sewage. Even if none of them said a word, the youthful crowd alone could intimidate the Regional Water Quality Control Board with a powerful message—don't irritate future voters.

There was one catch: the hearing was scheduled for a Monday morning. Granted, this might have been an additional incentive for the students—get a civics lesson and cut classes at the same time—but Bennett would have to convince the school administration that the event's educational value outweighed the apparent invitation for truancy. Then again, he didn't actually go to his principal, Glen Cook, and ask permission to take the day off for himself and a couple hundred students. No matter. Mr. Bennett was a minor celebrity among the two thousand students at Culver City High School. They had pinned up newspaper articles about him on hallway bulletin boards and watched him on television. "It wasn't a secret what I was trying to do," he says now.

So instead of seeking school sanction for bringing the students to the meeting, Bennett simply made it clear he was going and all were invited. This might have been a tad presumptuous of him, but, according to Bennett, Cook didn't say anything to suggest it was a problem. To at least give the flavor of education to the meeting, Bennett convinced two social studies teachers this would make a great civics-lesson field trip, and they arranged to bring their classes too.

Despite the wink-wink nature of all this, Bennett still figured it was only fair that the school chip in for buses to take the anticipated teenage hordes from the school to the state building in downtown Los Angeles, some twenty miles away, where the hearings were to be held. Principal Cook said no. Bennett says he smiled and calmly walked out of Cook's office, but the denial chapped his hide just the same. As far as he was concerned, the school seemingly had plenty of money for buses to take sports teams to competitions, which everyone deemed an important activity, and yet the school couldn't scrape together the cash for transportation to an event that would expose the kids to how the real world actually operated.

Bennett figured such misplaced priorities could be overcome and did what he always did—assumed that those in authority were, shall we say, misinformed, and took his cause to the people. In this case, that was the student body council, which, as it turned out, was sitting on a few bucks. He appealed to each class council, sophomore through senior, and they all agreed to, in essence, override those in charge and pay the buses' operating expenses out of their own budgets. No bake sale necessary.

To be sure, this wasn't the only guaranteed crowd for the hearing. Dorothy Green put out the call for her friends and co-activists to not only attend but also put in a word or two for the cause. Also, on the day before the hearing, the *Los Angeles Times* published an editorial in its Sunday edition, written by Patrick Wall, executive director of Earth Alert!, which laid out a fairly calm but gruesome description of current conditions in Santa Monica Bay. He mounted a brief attack on SCCWRP by asserting that the research facility "has 'demonstrated' more than once that its sewage actually was beneficial to the area's sea life. . . . In fact, this blind watchdog has even advocated the disposal of sewage sludge—the solid matter left after treatment—in coastal waters, which is illegal under the federal Clean Water Act."

Interestingly, Wall claims the *Times* fact-checked the article from every angle, but the editors missed one arcane but important mistake: "Until now," he wrote, "Hyperion has had no problem getting this [301(h)] waiver." Oops. The waiver had been pending for three years and was never actually granted to the city. Wall ended his editorial with an invitation to the paper's million-plus readership: "The public now has a rare chance to stop this poisoning of our coastal water. If you love this ocean, please testify on its behalf." If, as Bennett claimed, the EPA and water board had furtively announced the first hearing, the second one was no secret.

On May 13, Bennett, dressed in a dark suit, boarded each school bus before it left the campus and exhorted his charges (who were escorted by volunteer parents; all of this Bennett arranged himself) to behave themselves. This meant no booing or cheering or otherwise behaving as rowdily as he thought they would if left to their own consciences. He then drove with Bente to the hearing in their secondhand Chevy Impala and walked into the state building's room 1138, a somewhat drab, flatly lit auditorium with theater-style seating for the public and tiered rows in the front facing the audience where the six-member board sat behind a waist-high partition (there were supposed to be nine on the panel, but three positions were vacant). Bennett took a seat in the front row, while Bente found a spot in the back, deliberately away from the action and the spotlight about to hit her husband.

"I've had threats on my life," Bennett says now. "I didn't want her to be involved at all. I can't tell you, you know? Gee whiz. She'd come to all the demonstrations, but I never introduced her as my wife. She'd come to all the meetings, but I didn't want her to get involved, obviously involved." He didn't carry it to the meeting, but after an unspecified number of phone threats, Bennett had bought a Smith and Wesson .38 handgun for protection.

Students, their parents, environmentalists, city and county officials, and others quickly filled the chamber. Bennett's tactic for demonstrating public outrage had worked so well that boisterous warm bodies overflowed into the hallway. More important, forty-eight people signed up to testify, nearly all of them against the waiver. But the true civics lesson started as soon as the television news videographers showed up and planted themselves at the left side of the room near the front. That's right, kids, anyone there could have told them; it isn't enough to tell government officials you think they're screwing up. You have to do it front of the media. It's that symbiotic relationship between protester and reporter that gets the most done. And Bennett had assured every media outlet in Southern California that the hearings would attract a newsworthy crowd.

Rim Fay sat next to Bennett grinning. For years, he had spoken in nearly empty rooms to largely unimpressed decision makers, and if his bitter tone at the previous, March 25, meeting was any indication, the unrequited effort had worn him down. But now! On this day, hundreds of indignant people embraced his beloved bay. The *Los Angeles Times* estimated that two hundred people filled the room. Patrick Wall, who arrived late and couldn't get a seat, was stuck in the hall—with all the other elbows and shoulders knocking together—while trying to keep track of the speakers.

Several rows away from Fay sat SCCWRP's Willard Bascom, looking unmoved by the attention given the 301(h) waiver.

The EPA's Eklund, looking almost prim in her beige suit coat, midlength blonde hair, and chunky necklace, passed a crowd of protesters in the hall on her way to the meeting room, a little intimidated at first by the scene. But she reminded herself to stay calm, and by the time she opened the hearing, she appeared professional and objective. She began with a pro forma recitation of the 301(h) waiver's restrictions, sounding as though she wanted to make sure the antiwaiver crowd realized the law wasn't the free ride Bennett and others would have them believe. The city still had rules to follow even if they got the waiver, she said. Finally, to add a we-really-care-what-you-think tone, she looked at the crowd—the size of which the water board had never seen before at its meetings—and assured them, "We will stay as late as it takes to hear everybody."

Any goodwill that this might have created quickly disintegrated the moment board chairman James Grossman announced that the EPA's Sheila Wiegman and other government representatives would get the first crack at testimony. Bennett erupted from his chair, incensed his flock would have to sit through what he considered self-serving, ad nauseum speeches.

"I'm Howard Bennett. I'd like to—"

Grossman calmly interrupted him. "Mr. Bennett, I'm sorry, this is not the appropriate time to be heard."

"From what I understand we only have five minutes of public comment at the start to discuss what's not on the agenda—"

"Mr. Bennett, this is not the time to be heard. Mr. Bennett—"

"This board—"

"Mr. Bennett, I'd like to ask you to behave, as a favor, to conduct yourself in an orderly way. This is not a normal hearing—"

"This was supposed to be specifically for public comment—"

"You will have plenty of chances—"

"The public—"

Grossman sounded flustered. "I believe, Mr. Bennett, the media is getting a good opportunity to see you," he said, unable to resist the sarcastic reproach. "You will be on television very soon, Mr. Bennett—"

"This is specifically for public comment! And you're taking time from—"

"You will have a chance," Grossman said. "You will have a chance."

The audience applauded as the two continued sparring far beyond what could be considered reasonable debate.

Bennett wouldn't give up. "They've been heard before," he screeched, referring to the government officials. "This is specifically for *public* comment!"

Rim Fay charged to the microphone and, in his booming voice, unleashed his customary witness speech: "I've been attending meetings with this board for over twenty years, more than any board member present, probably more than any person present in the audience today. This meeting was continued to hear from the public! There is no law, there's no regulation that says you have to hear from governmental agencies before you hear from the public. I would suggest that you hear the presentations from the public first. . . . We request this board to listen to them!" The crowd loudly applauded his populist exhortation as though happy someone had introduced this us-versus-them sentiment so early in the hearing. The cheers encouraged Fay to bellow, "And then the governmental agencies which are legally responsible for the problems which bring us here today, and have not stopped those problems for more than forty years—" More applause cut him off before the statement developed full coherency.

Board member Betty Werthman lectured a bit about decorum, and finally, another board member, Celso Martinez, shut down what was becoming a circular argument by calling for a vote on this issue. They defeated the motion to have the public speak first, but Bennett tried one more outburst. This time, Grossman threatened to kick him out of the room. Bennett and Fay finally sat down.

"I was a pretty uppity fellow," Bennett says now in the kind of cheeky tone you might expect from a teenager. He adds that his blood still boils when he thinks about the board's insolence—as he sees it—toward the public.

Among those representing the city was Dr. John Dorsey, the marine biologist-slash-surfer, who wanted people to believe he was presenting an evenhanded, objective, and scientific point of view. And indeed, he conceded up front that the city should not be dumping sludge from the seven-mile pipe: "The first process is going to be removing sludge from the bay. I feel that is a really important step and, I think, a very major step for the City of Los Angeles."

Dorsey continued, outlining the teeming life near the five-mile outfall where the primary treated sewage landed. "Out of one of those grabs in that area," he said, referring to samples he had taken, "we get up to a hundred species of organisms, with nearly 35 to 4,000 individuals in a single grab. And on the average, we'll pick up anywhere from 50 to 60 species of organisms with several thousand individuals. We have a lot of biomass in that area and diversity." Obviously, the animals were small—a thousand snails less than half a millimeter in size and nearly 150 clams less than five millimeters in size. Later, in a statement seemingly directed at Bennett, Dorsey said, "In summary, I want to emphasize it seems like this has been a very emotional issue. Over the past weeks, we have been accused of discharging raw sewage into the worst polluted bay on earth, resulting in death zones. . . . These statements are probably based on hear-say and not a lot of facts."

Dorsey doesn't remember much about the hearing today, except for the relief he felt when he finally finished his presentation. "All I know was it was a big circus," he says now. "I was glad to get out of there."

After forty-two minutes of such government testimony, Bennett got his chance to spit out the kind of flamethrower oration he had honed over the past several weeks. His words so effectively built upon the *ick* factor intertwined with the topic that, at lunchtime, no doubt few people visited the hotdog stand on the street.

He started by expertly citing unattributed facts with pure emotion: "Because of sewage dumping, most of the Santa Monica

Bay kelp beds are gone. . . . Our fish have been poisoned so that we're told not to eat them now. I don't mean to be strong and emotional but, frankly, anyone who doesn't think we have a problem is either near-sighted, blind or unfortunately, on some days, has lost their sense of smell.

"The general public has been treated like a child," he continued earnestly, "and has been given a false sense of security by the entire sewage disposal industry. As an English teacher it is clear to me that the sewage industry uses euphemisms to pacify the public and fool them into thinking they're doing a good job. Euphemisms are the use of good words for evil words. A child is told that its mother or father are now 'at rest' when in fact they are dead. The sewage industry talks about its 'primary treatment process.' This is another euphemism. It is separation of the big pieces and not treatment at all. The general public is not to be fooled like a child. They understand that what is left after separation is still raw sewage. To use the sewage industry's euphemisms, the type of 'waste water' or 'effluent' that is discharged after 'primary treatment' is what can happen when we get a 'a gastro-intestinal upset,' or as the public would say, just plain diarrhea. No big pieces, but raw just the same."

Granted, this dissertation strayed somewhat from the factual portion of his testimony, and no doubt out of good taste the television reporters didn't choose it for their sound bites. But this was pure Howard Bennett, distilling the topic to the kind of gut-level imagery people grasped first, long before the facts ever muddied the discussion.

Bennett also understood his ultimate audience, the board members. While Betty Werthman rocked back and forth in her executive chair and others leaned against the table in front of them expressionless, he said, as if warning them, "I have not come here alone today." The crowd interrupted him with applause that lasted thirty seconds. "As a teacher at Culver City High School, the students wanted to be here. Representatives were chosen and have come. . . . As the voters of tomorrow they want to see this permit denied and be able to swim, fish, and surf safely in the waters of Santa Monica Bay." When Bennett finished his oratory fourteen minutes later, the group rewarded him with a friendly, back-slapping applause, as though these were all his friends showing support.

The speakers kept coming throughout the morning, including one of the current heavy hitters of the enviro set, Carla Bard, former chairperson of the State Water Resources Control Board, and several scientists who clung to the gray areas within the debate, making it hard to know which side they supported. Occasionally, acting like a ringmaster, Bennett would ask the board to indulge him by allowing one speaker or another—each who needed to leave soon—the opportunity to leapfrog ahead on the speaker list and testify next.

Altogether, four Culver City High School students testified, including one who claimed a myriad of ailments he attributed to ocean pollution: severe chronic infections, sinus growths, and the need for two operations. "I think the water is polluted, absolutely, and you should do something about it," he concluded with a slight adolescent twang.

About three hours into the meeting, Dorothy Green took her written statement over to Felicia Marcus, a woman she had only just met in person, despite their many phone conversations regarding water issues, and asked her to read the testimony; she had to leave to pick up a friend's children from their babysitter.

Marcus gamely presented Green's speech, which tried to tie the 301(h) waiver to overall water usage. The slightly puzzled audience politely clapped at the end, but the meaning was unclear.

Finally, nearly five hours into the hearing, Rim Fay's resonant voice filled the room with a rambling, scolding tirade directed at the board. "I think there's a problem," he said, "with the way this hearing started off with a distinct bias demonstrated by the chairman, where the chairman was concerned with the clientele of this board which this board licenses to pollute." The insults continued throughout his seventeen-minute speech, which was part snide review of what the board had done and part gruff appeal for them to deny the waiver. He finally ended in a crescendo of words that banged home a theme his friend Don May had espoused—a clean bay meant more dollars for fishing and tourism businesses. Unlike all other antiwaiver speakers, he went back to his seat with no applause.

In a way, Fay didn't do the waiver opposition any favors. As Robert Ghirelli, the board's executive officer at the time, puts it now, "He was kind of aggressive, acerbic, and attacking in his style. While I was a member of the staff—not a member of the board itself—that kind of style . . . it usually doesn't go over very well in a setting like that. The board gets turned off with what people are saying because they're paying [more] attention to the style, the way the message is being delivered, than they are to the message."

SCCWRP's Dr. David Brown sat in the audience listening to other scientists tell the board the pollution horror stories he wanted to confirm, but couldn't. He also heard other scientists dispassionately try to show how full secondary treatment wouldn't cleanse the bay as some believed. He wanted to dispute that, but couldn't. Instead, he sat still as the board promised to consider the 301(h) waiver at its next meeting, on July 22, skeptical they would do anything but approve the waiver. But again, he couldn't publicly object. Brown had been ordered by his SCCWRP boss, Willard Bascom, to say nothing.

CHAPTER 9

The Scientist

The silence Willard Bascom imposed on Dr. David Brown lasted no more than four days. The following Friday, May 17, at another public hearing, Brown publicly accused his boss of laundering SCCWRP's research into bright, stainless conclusions instead of allowing his scientists the freedom to voice their belief that Santa Monica Bay was so polluted that its diverse marine life had largely disappeared in places. And while the statement itself was fairly brief, Brown figured this should have ignited the city—should have inflamed environmentalists and the thus-far laissez faire politicians alike. He couldn't imagine riots in the streets, but those would have been nice, too.

Unfortunately, the one television news reporter present—working for KNBC-TV—walked out of the room apparently unimpressed. Brown, who thought his revelations bashed just about any other news for the day, felt perplexed at the snub but, at the same time, resigned to the idea that the information wasn't considered all that appalling. This is not to say the press

Dave Brown in 2008. The scientist blew the whistle on Willard
Bascom's rosy, distorted reports on Santa Monica Bay's health.

completely ignored him. The next day, the *Santa Monica Daily
Breeze* ran a story by Anne Morgenthaler breathlessly titled
"DDT Research Ordered Hushed?" but the city yawned. No
other reporters called Brown for a quote or to dig a little deeper.
Here he was, screaming "scandal" out loud, and the media gatekeepers treated it as though he had announced the church
picnic.

Rim Fay, who had been encouraging Brown to publicly strip
Bascom naked of his pretense, happily called him and cried into

the phone, "You did it!" as though he could see his nemesis about to fall. "You did it!"

But Brown felt less triumphant. If that one newspaper article had failed to resonate beyond just a few people, then he needed to repeat his show of frustration—no, anger—over Bascom's behavior again and again. So he next chose what seemed at the time a reasonable, though possibly passive, gesture—he wrote a letter to state assemblyman Tom Hayden, the man to whom he had testified on that Friday about how Bascom had told his entire staff in 1984 to keep their mouths shut over the bay's pollution.

Brown read the five-page letter after it clattered off the dot matrix printer. After what had happened with the press, he figured the missive would have the half-life of a struck match.

It seemed so passive a gesture. Sure, he destroyed Bascom's contentions, point by point, that the bay's marine life was happy and well fed by the sewage pouring out five miles offshore. (This was such a bizarre notion that its hokum quality should have been obvious to all, but Bascom said it so credibly, people believed it without judgment.) He also revealed that Bascom had obfuscated the data on the cancer risk from eating DDT-tainted fish. This was white-hot, this was radioactive, this was *important*. But sitting on the page in that blurry dot matrix print, it seemed incapable of stirring anyone. He handed the letter to his wife, Anne, and asked, "Well, do you think I should send it?"

Anne knew the letter could get him fired, and perhaps that should have worried her. They had been married only three months and had just bought their small, three-bedroom Long Beach house. The two didn't have a lot of money: she was a USC graduate student with a fifteen-thousand-dollar-a-year internship, and a good chunk of his paycheck went to his ex-wife in Vancouver, B.C., for child support. Even though Dave managed money better than most people, if he lost his job at the Southern California Coastal Water Research Project for outing his boss, they would be stepping into deep sludge. Nevertheless, she left much of this unsaid. "I totally trusted him and had confidence, and, okay, this is important to him therefore he needs to do it," she says now.

The day before, on Friday, Brown and Bascom had sat in Santa Monica City Council chambers, at a hearing held by the State Assembly Task Force. It was chaired by Hayden, who by this point had sunk his teeth into the Santa Monica Bay issue with the kind of combined enthusiasm and publicly expressed outrage that could come only from a politician eager for a righteous fight. Bascom did most of the talking, perhaps realizing that Brown, his director of chemistry programs, wanted badly to tell Hayden the truth about SCCWRP's research.

Bascom's filibuster nearly filled the 11:30 to 1:00 time slot allotted to him, Brown, and Fay. The three had been asked to present evidence of toxic contamination in Santa Monica Bay's water and fish, as well as discuss the health risks to humans. Bascom came loaded with charts showing how DDT and PCB emissions from Hyperion and Los Angeles County's Carson Sewage Treatment Plant had declined from extremely high points in 1971 to barely measurable low points in 1984. The two compounds, he claimed, had also decreased in marine life, from about 18 parts per million in 1971 to barely traceable levels in 1981. He told the panel that, of all the sport fish collected from the metropolitan Los Angeles area, none had exceeded the EPA's recommended toxicity standards of less than five parts per million for DDT and two parts per million for PCBs.

Bascom cheerfully compared the health risk of eating white croaker caught in the bay to that of dying in a car accident: you have a 1-in-70 chance of dying on the road, he said, while your risk of getting cancer from eating the fish was 1 in 13,888. As Brown remembers it, Hayden acidly asked, "Is that the good news or the bad news?"

Bascom even threw in at the end his pet notion that the organic material pouring from Hyperion's outfall actually provided one huge underwater dining hall for fish and other marine animals, so numerous now that one could only conclude that sewage was actually good for the biomass.

The barrage of what Brown considered misleading information continued while he sat nearby, twitching in his chair. He knew the truth! He was prepared to fill in the missing pieces that Bascom deliberately left out—the complete story that would show that there were a lot of fish but only a few species. The rest had vacated the area, leaving a bay devoid of diversity and, therefore, an unhealthy ecosystem. And the cancer risk from eating fish caught in the bay was far higher than Bascom had stated. Bascom based his numbers on people merely nibbling a few grams of white croaker, not the four to fifteen times that amount actually consumed by Southern California sportfishers.

Finally, just before one o'clock, Hayden called Brown to the front of the chambers and asked him, in a seeming non sequitur, "Has the SCCWRP staff been pressured to keep any information from the public?" Brown's five years at SCCWRP could have ended right there.

Brown was a precise, methodical man, and knowing the truth but not being able to scream it out frustrated him. The careful research he had done went into SCCWRP's annual reports, and for anyone willing to spend a few hours choking on scientific vernacular, the truth was certainly there. Scientists elsewhere sought out SCCWRP's research, recognizing how thorough and accurate its data were. And they certainly recognized what the dense collection of charts and graphs whispered about the bay's pollution. But none of them saw it as their job to publicly mention the gap between the dour data and Bascom's optimistic interpretations.

Apparently, politicians' staffs, city and county engineers, and overworked bureaucrats rarely if ever cracked open the reports for any longer than it took to read Bascom's summaries of the research. There they got the message that, even though the once pristine bay was now hurting, things were getting better. In the readers' minds, this equated to: We must be doing the right thing. No need to spend any more money than we are now. Surely we qualify for the 301(h) waiver!

Just as aggravating to Brown was how Bascom had turned Brown's research into an excuse to continue the pollution. Bascom had used him.

Back in 1980, when he joined SCCWRP, Brown was a marine biology whiz kid. After graduating with a bachelor's degree in biochemistry from the University of British Columbia in 1970, he immediately went to work for BC Research, a research and development company that, at the time, was working on a government-funded project to study how pulp mill wastewater had affected the Strait of Georgia's marine life. He learned so much about real-life research methods that, when he went back to school five years later to get his doctorate, he banged out twelve published papers during his graduate student tenure, a better showing than that of most of his professors.

For his PhD, Brown had studied marine organisms' ability to ingest the various toxic pollutants spicing their food and then detoxify—or neutralize—those same contaminants so that, presumably, they have little effect. Metals are detoxified as they bind to a protein called metallothionein, while organic toxicants are metabolized and bound to glutathione, in essence detouring the toxicants before they can poison the animal. Most animals possess this ability, but with one caveat: they can't detoxify all the toxicants they ingest. And strangely, it appears that, although there is an upper limit to detoxification where acute toxic effects occur, even at lower levels these detoxification systems miss some toxicants. In essence this means that, if the amount of toxicants doubles, the amount the body can't detoxify also doubles.

This is an important point. If there is just a small amount of, say, DDT in the environment, then it probably can't harm the organisms that ingest it, even when they can't detoxify all of it. But the higher the DDT levels, the more remains in the animals' fatty tissues. And the problem gets worse as big fish eat smaller fish: the relatively minor concentrations of toxins in the small fish become magnified tenfold at each step as the toxins make their way up the food chain. It's a little like drinking a glass of wine of which your body metabolizes only a portion of the alcohol; the rest is stored away. At your next meal, you drink another glass and the same thing happens. Pretty soon, your insides are swimming in booze. In the case of really big fish eating the average big fish, the concentrations of toxins are many, many times the concentrations found in the tiny fish who originally took in the DDT or PCBs.

As near as Brown could tell, no one else had researched detoxification in marine animals, and so the Scripps Research Institute in La Jolla, California, took an interest in him. He drove from Vancouver to La Jolla and interviewed with the prestigious group, and they offered him a seventeen-thousand-dollar annual salary, which might have been enough for a single guy willing to room with three others. But Brown wanted to have enough money to fly his beloved sons, who were six and ten, down for extended visits and for him to go up north to see them.

Scripps's director suggested that a friend of his, Willard Bascom, at SCCWRP, might also be interested in hearing about detoxification. It's a rare scientist who doesn't want to talk about his research pursuits, and Brown dropped in at SCCWRP's Long Beach offices on his way back to British Columbia.

For thirty minutes, Brown's ego bathed in Bascom's eager attention to his discussion of detoxification. What a wonderful word, *detoxification!* There was something so magical about it. The idea presented this bold image of near perfect purity even when horrible things ended up in a critter's environment. The word was at once scientific-sounding and easily understandable to the laypeople with whom Bascom dealt. Who would dare deny the 301(h) waiver to Los Angeles city and county when God's creatures could *detoxify?*

This thirty-one-year-old kid with the thick blond hair and neatly trimmed beard who looked suitably academic enthralled the former adventurer, the man who probably thought he had seen it all. After only half an hour, Bascom offered Brown a job at SCCWRP, one that would pay barely enough—$27,000 a year. However, given his financial acumen, Brown could stretch it as far as he needed, and, more important, the salary would allow him to see his sons, whom he missed and hated to leave.

Brown became Bascom's star employee. While Bascom often bullied others, grinding down even some of his best scientists to the point that they quit, Brown escaped the harassment. He happily went about his research, almost innocently fascinated with studying detoxification right down to the molecular level.

But even as buried in his research as he was, Brown soon recognized how polluted Santa Monica Bay was, and that detoxification didn't solve everything. Trace metals, organic toxicants, and solids escaping the primary sewage treatment caused changes at the food chain's bottom rungs, in the phytoplankton and diatoms, which the smallest of fish and other animals ate. The fish were picky eaters, so sensitive that they could discern when their food wasn't quite the same. They consequently left the area for locations offering just the right kind of dining opportunity. Forget about detoxification. When the small fish left, so did most of their predators. And soon, only a few species—those who could tolerate what was left—stayed behind to proliferate.

But Bascom didn't publicly reveal such details. And after a couple of years, Brown soon realized that Bascom's real mission wasn't to bring out the truth, but to bring home the 301(h) waiver by proving that full secondary treatment wasn't necessary. He endeavored to prove it in a number of ways, and one of them was by touting detoxification.

Brown watched Bascom slide around the outskirts of the truth, seeming to understand that people usually didn't question Bascom's authority as SCCWRP's director, and that they accepted his conclusions. At one point, Brown found a copy of

Bascom's testimony before Congress's Subcommittee on Water Resources on May 24, 1978, where justifying the 301(h) waiver was on the congresspeople's minds. "Dr. Bascom," as the transcript called him (notwithstanding the fact that Bascom, an adjunct professor of ocean science for the University of California, San Diego, hadn't graduated from college), claimed that, on "the California coast, the effects of waste in the past have been shown to be not very long lasting and quite reversible; if we look at outfalls that were once used and are now abandoned, they reverted, in most cases within a year or so, to the original condition, whatever that was. . . . And the situation around all the existing outfalls is steadily improving anyway—or at least, almost all of them; there might be an exception [nowhere does he note what that might be]. This means there is no real concern that the situation is going to get greatly worse while we are doing our homework. Things are getting better anyhow."

It was a remarkably slippery statement—if someone bothered to parse the words—but the congresspeople were no doubt so reassured that they didn't stop to ask for details. However, the truth was that the sewage dischargers for whom SCCWRP worked had no plans to abandon their outfalls, and so to even discuss this was a typical Bascom red herring. The second part of his statement was essentially false; things were not getting better.

Later in his testimony, Bascom tossed them his favorite justification for the 301(h) waiver—that sewage was good for the fish. "What the outfall does," he said, "is put additional organic matter on the sea bottom, and the animals that live off that tend to congregate there. . . . So you will see [referring to a chart] that, as the diversity of animals decreases, the number of animals goes up. Instead of having a few animals of a lot of species, you have fewer kinds of animals, but you have a heck of a lot more of them. Now, I have no idea whether that is better or worse to have it that way. I do not think any man can ever answer that question." In fact, most marine biologists considered lack of diversity the sign of a sick environment.

Around 1982, Brown watched one night while Bascom told a television reporter that SCCWRP couldn't pinpoint where the DDT known to be in the bay had come from. "My guess," he said, "is it could have drifted across from Japan, Korea or the Philippines."

"Bascom knows where it came from!" Brown yelled at the television. "We all know! It came from LA County's sewage treatment plant." As far as Brown was concerned, Bascom was purposely misleading the public, knowing that SCCWRP's own research—as well as that of other scientists—had shown that DDT and PCBs had peaked in the sediments and animals at Palos Verdes, where the county's sewage treatment plant discharged its effluent. The hundreds of tons of chemicals were so pervasive they could be found all the way to Oregon and to the tip of Baja California. Brown was so incensed by Bascom's duplicity that he considers this his "turning point." In a short time, he would expose his boss's lies.

It's possible that Bascom thought he had a justifiable rationale for obfuscating the truth. Both Brown and another SCCWRP scientist, Dr. Bruce Thompson, recall Bascom feeling the heat from Charles Carry, the county sanitation district's chief engineer and general manager. Without ever saying so directly, Carry left the impression in many staffers' minds that if the county didn't get its waiver, SCCWRP's funding would dry up.

"I do know there were times where there were those kinds of veiled threats because of the waiver scares back there," Thompson says. "And while I actually never heard Chuck Carry say those things, it was reported to me by others that he had made those threats about [a] 'remember where your money's coming from' kind of thing."

Robert Miele, who worked under Carry as the district's head of technical services, says Carry could be imposing at times, but doesn't think Bascom would have blinked at any threats, implied or otherwise. "If you met Willard, you'd have a hard time believing that anybody could pressure Willard to do anything. He was truly a wild hair." And later he adds, "[Carry] had a real abiding pride that SCCWRP was this independent group, so it would be hard for me to believe, knowing Chuck, that he would then get into the position of saying, wait a minute, we're going to play politics here. You're not helping us, and so therefore stop doing it."

Nevertheless, Brown remembers Bascom saying on several occasions, "Don't bite the hand that feeds you." And everyone knew what that meant.

Just the same, Brown attempted at least a nibble. He first tried a passive-aggressive gesture during a public meeting in May 1984, where SCCWRP scientists gave biannual reports on their research to its funding agencies and the public. Bascom knew that among the few people attending the presentation would be Rim Fay and his friend Don May of Friends of the Earth, so he required anyone giving a report to first rehearse it before him to ensure that its content didn't stray beyond certain boundaries of candor. Brown wanted to say out loud that, in their study of DDT and PCBs in the bay, they had gone as far as

ninety miles offshore to try to find control fish without the toxins (that is, specimens to compare against the tainted fish), but
couldn't. Bascom even said as much in a 1982 SCCWRP paper,
but he blurred the specifics: "So began a search for truly accept-
able control animals—but even after sampling at such distant
locations as San Clemente Island and Cortes Bank in the southern California Bight; Punta Banda, Mexico; and Morro Bay on
the open coast north of Point Conception, we still do not have
uncontaminated samples."

Instead, the fish contained four to nine times as much DDT
and PCBs in their systems as the fish swimming at that time in
Commencement Bay at Tacoma, Washington, which had been
declared a Superfund site the year before because of its widespread contamination from local industry. (When the EPA
designates a Superfund site, most often those responsible for
the pollution are required to clean it up. The term has also
come to mean any place so defiled by deadly toxins as to be
harmful to all life.)

Bascom deleted the Superfund reference, claiming it was too
political. As near as Brown could tell, that referred to Bascom's
contention that a local angry scientist had exaggerated Commencement Bay's pollution. Therefore, a comparison between
the toxicity levels in Santa Monica Bay and this supposedly er-
roneously designated Superfund site was a spurious comparison. However, Bascom went further and in SCCWRP's 1983–
84 biannual report wrote, "These data are difficult to interpret
because of inadequate controls." That is, his researchers couldn't
find any fish clean of DDT.

Brown figured there was another explanation: If he said out
loud that the contamination levels in fish in Santa Monica Bay

ninety miles out were worse than in fish at an established Superfund site—and that the levels of toxicants increased closer to shore—then the EPA would finally recognize the pollution's extent and deny the 301(h) waiver.

(Unknown to Brown, an EPA official had a "gut feeling" that SCCWRP's overly optimistic conclusions were tainted. However, this person says now that because of heavy workloads nothing was said, and lack of proof squelched any desire to pursue the matter.)

It's not that Fay and May didn't already know about the latest DDT research; Brown had secretly told them. (Fay and Brown had met through what was considered a normal exchange of information—that is, Fay had probably called Brown to ask about his research.) But if Fay and May had simply repeated the data, its credibility could have been questioned because it didn't come directly from a theoretically neutral—or at least more reputable—SCCWRP scientist. After all, Fay had been talking about the near apocalyptic conditions for years—even suggesting the bay be designated a Superfund site—without stirring much interest. This might have been because he had a habit of calling himself a "simple fisherman" and this kind of modesty backfired. (It led some marine biologists to believe his conclusions were based on observations alone and not full-blown scientific studies and, thus, were suspect.) Or perhaps people—particularly the decision makers—just got tired of hearing him. This may also be why Howard Bennett later succeeded where Fay hadn't. In some ways, Bennett was even more obnoxious and less insistent on scientific details; nevertheless he was a new face with a different approach to the issue. Sometimes, that's all it takes.

Recognizing that when all else fails, you must get a reporter to stir the pot, Fay invited the *Los Angeles Times* staff writer Richard O'Reilly to the presentations at the May 1984 meeting. Brown knew O'Reilly was coming (but Bascom didn't), and he welcomed the opportunity to publicize his research even if it was slightly expurgated. The *Times'* headline on the next day, May 16, read, "Pollution along Coast Surprises Scientists." The article quoted Brown detailing how his staff had tried without success to find uncontaminated control fish in a 300-mile swath from Port San Luis (about 135 miles north of the bay) to Ensenada, in northern Baja California, Mexico, and as far as 90 miles offshore. O'Reilly, left uninformed about the Superfund comparison, reported only on Brown's watered-down version of the study.

Not watered down enough for Bascom, who hauled Brown into his office to discuss the overly enthusiastic candor. Brown, in the first of many disingenuous explanations, told Bascom that all he was trying to do was present the facts as he knew them, which was—as he saw it—his job. Bascom accepted this but counseled Brown to be a little more circumspect in a reporter's presence.

Publicly, Bascom soft-pedaled Brown's report by repeating his usual vague contention that the ocean water quality was "steadily getting better all the time." O'Reilly further quoted him as saying, "We're pretty sure there's no human health effect, and we can't detect any fish population changes." Given Bascom's position as Brown's boss, his contrarian outlook effectively deflated the story. Even if readers were apt to be horrified by so much DDT swimming about the ocean, the article came out weak. It appeared by the last paragraph that things weren't really so bad after all.

However, it still remained that the fish in the bay were virtually all tainted with DDT and PCBs, and therefore the compounds were finding their way to humans who fished and ate white croakers, among other catches.

After the article appeared, Bascom went further on the offensive. In an editorial for *Sea Technology* magazine in June 1984, he took advantage of Brown's detoxification research: "Although low levels of [DDT and PCBs] are widely dispersed in the ocean, their actual toxic effect is small because natural detoxification mechanisms built into all animals cause these chemicals to be metabolized and the products made unavailable by a sub-protein called glutathione. Thus sea animals and the persons who eat them are both protected." He also, again, touted sewage's supposed nutritional feature when he asked, "Is it conceivable that there is something wrong with putting food and nutrients in the ocean?"

A month later, the *Los Angeles Times*, doing the best it could to inform the city of the bay's pollution, pitted Bascom against Rim Fay in a faux debate where their quotes were juxtaposed, but the two men—no doubt by choice, given the animosity between them—were never in the same room while being interviewed. The article portrayed the two as dismissing each other as crackpots. Fay came off as somewhat brooding: "I saw animals dying on the ocean floor," he said of his diving experiences, "rotting in their shells, huge areas that have never recovered." Bascom cheerfully called Santa Monica Bay "a marine garden."

This kind of evenhanded approach, while laudable from a journalistic ethics standpoint, failed to arouse much interest. And, just to keep things in perspective, almost an entire year would pass before Howard Bennett—who didn't have a newspaper subscription and was largely unaware of this—would learn of the 301(h) waiver. However, when he did find out about it, his ire was only indirectly a result of a news article. It was Rim Fay who actually lit the fire under Bennett.

It was hard to tell whether the city as a whole didn't care or whether, as many environmentalists believed, people simply thought DDT on the ocean floor didn't stop them from getting their kids to school or paying the bills. As ugly as the pollution was, day-to-day life still trumped everything else, including three-eyed croakers.

In early November 1984, recognizing his staff's unease, and in particular Brown's continued desire to get the facts—*all of them*—out to the general public, Bascom held a staff meeting to make sure they understood SCCWRP's true mission. "The main reason why the cities and counties of Southern California are willing to pay for our studies," he told them, "is because they are engaged in a long-term struggle with the EPA over whether secondary treatment is to be required of every discharger." He contended the SCCWRP scientists weren't muzzled and could present their findings or opinions, "so long as those were supported by adequate data that, after consideration, we as a group agreed on."

And finally, he got to the heart of his complaint. "We must try to understand the problems of the senior sponsors [that is, the heads of the waste dischargers]. The waiver of secondary treatment and accompanying monitoring is their main concern." He brought up a 301(h) waiver application denied in Puget Sound after public reaction to pollution. "But I would not like for us to trigger a public reaction here that would give EPA Region IX a *political* excuse for a similar ruling."

Bascom also called this meeting because several SCCWRP scientists had written a research paper regarding high levels of chlorinated hydrocarbons (DDT) in marine mammals, and Bascom wanted to quash its publication. We could lose our jobs, he said, if the information got out. With Bascom standing at the head of a conference table, Brown sitting at the other end, and all the staff scientists between them, Brown not so casually said, "I'm willing to lose my job." And each scientist in turn agreed to unemployment if it meant getting out the truth. "That was really a big deal," Brown says now, "because it really showed that Willard was in fact threatening the staff, and the staff was not cooperating."

Despite Bascom's attempts to downplay the DDT issue, in February 1985 a reporter for KCBS-TV, David Garcia, eagerly told the city that Montrose Chemical Company had *legally* dumped barrels of DDT and other chemicals off Catalina Island presumably because it was the most expedient way to get rid of what was considered waste material. In a sense, Garcia was justified in calling this an "exclusive," given the ho-hum reactions other similar DDT stories had received in the past. Just the same, Garcia, whose assistant told people at SCCWRP that the movie *The China Syndrome* had been fashioned on his experiences (presumably he wasn't played by the film's star, Jane Fonda), was mining old territory.

The story covered barrels deliberately dumped off boats between 1947 and 1981 by Montrose, the last U.S. manufacturer of DDT, and it stated that this had been done under a permit issued by the Regional Water Quality Control Board. There's no denying this was sobering news. Garcia seemed out of breath as he listed off the other chemicals that also went into the water—beryllium, formaldehyde, napalm, cyanide, acids, solvents, and unidentified pathology wastes. As a gruff Rim Fay said in Garcia's report, "This is probably the most extensive and severe pollution problem ever documented anywhere in the world ocean." This sounded suspiciously as though Fay was referring to all the DDT in the bay, not just that from the dumping, but Garcia didn't make that distinction.

Unfortunately, Garcia's revelation merely covered the nickel and dime stuff. Most of the DDT covering the bay's floor had come via Los Angeles County Sanitation District's sewage treatment plant (full secondary treatment would have kept much of it out of the water). For decades, Montrose had been allowed to flush DDT waste from its operation directly into the sewer system, which amounted to thousands of tons of the insecticide making its way into Santa Monica Bay. This famously affected the brown pelican, which nearly died out after the birds ate the fish that ate the fish that swallowed the DDT, and the chemical made the birds' eggshells so brittle that they kept breaking before anything hatched.

Garcia's competition, KABC-TV, quickly tried to deflate his story the next day by trotting out the ever-dependable Bascom, who dismissed this as old news contained in a twelve-year-old SCCWRP report. In a reassuring, almost fatherly voice, he said, "Nobody thinks it's a real good thing to put toxic waste in the ocean. The question is whether or not a reasonable possibility [exists] that the toxic level can rise high enough to where it causes any damage to either man or the sea animals. And from the measurements we have made of sea animals, [I] suggest that would be an extremely rare circumstance. Nothing is a zero risk in this world, but I would say the risks are minimal. Absolutely minimal."

The Regional Water Quality Control Board wasn't so sanguine, however, and brought Bascom in to explain why they hadn't heard about this before. Well, actually, you have, Bascom said. SCCWRP had published a report in 1973 that devoted eight pages to the dumping of DDT, but apparently no one paid much attention. "I would say [the pages] contained essentially everything in the CBS report that triggered all this material," he said without expression. He went on to say, "The evidence that you have to date is only that a lot of material was dumped. We see no clear evidence that it's done anything, and I think you have a story which is probably at least twelve years cold." Still, in retrospect, the KCBS stories, and those by other reporters that followed, again hammered the general populace with the news of just how bad the bay's pollution was. But the news also pointed out how apathetic the public still was toward the subject: the rioting in the streets that Bascom expected didn't happen.

Later the same February, as if he were just a scientist spelling out the facts, Brown spoke before Assemblyman Tom Hayden's Santa Monica Bay Revitalization Task Force and finally told the world that the bay's pollution was much worse than that of Puget Sound's Commencement Bay. He and a USC professor of pathology, Harold Puffer, had just released a study showing that the white croaker held megadoses of DDT in its fat tissues. The researchers had also interviewed one thousand fishermen and found that 10 percent of them ate half a pound or more per day of contaminated white croaker, Pacific mackerel, Pacific bonito, and queenfish. Puffer was asked if he thought the fish threatened human health, and he replied, "Let's put it this way: I wouldn't eat it."

The next month, March 1985, Hayden again spotlighted the DDT contamination outlined in a Regional Water Quality Control Board report. "The staff report details decades of systematic neglect by the board in fulfilling their most basic functions," he told the *Los Angeles Times*. "The board appears to have been asleep at the switch, with the result being irreversible damage to the marine environment." He didn't know it, but his reprimand— unique among politicians of the day—was loud enough that it would end up in the hands of an old fisherman the next day, March 28, just as Howard Bennett was about to go for his morning swim. "Water poison!" the man would tell him, not entirely overstating the issue. It could also be argued that, had Brown not hammered continuously on the DDT, PCB, and sludge issues, the topic might not have gotten this far, and Bennett would still be blissfully swimming along the beach at Playa del Rey.

Hayden convened another hearing in May, a week after the second water board hearing on the 301(h) waiver. According to the invitation Hayden's office sent Brown, "The task force requests your appearance to testify on issues related to your agency's expertise and authority related to toxic pollutants in Santa Monica Bay." This was a little disingenuous on Hayden's part. Brown had already told one of Hayden's aides about the previous year's meeting where Bascom had threatened them with losing their jobs if they continued to push full disclosure and disregard the dischargers' desire for a 301(h) waiver. Perhaps wanting to create a little political theater, Hayden really wanted to ask Brown about that meeting.

So on May 17, after Bascom had finished his windy presentation, Brown stood before Hayden knowing that he was about to publicly reveal his boss's call for silence. All that time spent

angry, resentful over Bascom clouding the facts to suit the dischargers, was about to end. He could lose his job. He could lose
his career. All for a few moments of righteousness. His heart
thundered. His voice quivered.

"No need to be nervous, Dr. Brown," Hayden said.

Brown told the task force about the November 1984 meeting.
But that was as far as he got. Incredibly, Hayden adjourned the
hearing, and Bascom's presentation still sat there in the public
record, claiming that eating the fish in Santa Monica Bay wasn't
unhealthy.

The next morning, Brown wrote his letter. "Presentations
like Mr. Bascom's leave everyone confused," he told Hayden,
and then he methodically and precisely tore apart his boss's testimony, sparing no detail. He was direct in his critique. He
could have been a little more delicate, but he was angry, and
sometimes that's the only emotion that inspires honesty.

"Another thing Bascom didn't tell you yesterday," Brown
continued, "was that the fish in the photograph of the trawl
catch he showed you from the seven-mile outfall in Santa Monica Bay were almost entirely white croakers. Here we arrive at
the heart of the human health problem. As you know, white
croakers are the most contaminated sport fish in southern California. The reason why they are so contaminated is largely due
to the fact that they are attracted to sewage outfalls because
they feed on polychaete worms, which dominate the bottom-
living communities near outfalls. Since the outfalls are the source
of contaminants, animals which feed near them will be the most
contaminated. This is particularly bad news for sportfishermen
because white croakers are the most abundantly caught sport fish
in Southern California."

Brown also covered Bascom's misrepresentations involving the cancer risk to anyone who ate those fish. In fact, Brown said, the risk was two to three times higher than Bascom had told the task force. "It is important to realize that a large portion of the so-called sportfishermen in Southern California are not fishing for sport but rather as a source of food," Brown wrote. "Many of these people are unemployed and depend upon fish for a source of protein up to seven days a week. It is these consumers who are at the greatest risk in Southern California."

Once Anne Brown finished reading the letter, she thought, This is important. This is true. And besides, Dave was sending it to Tom Hayden, who would no doubt leave her husband out of it.

Dave addressed the envelope, and Sunday morning the two drove down Bellflower Street to a nearby mailbox in front of a Broadway department store. He held the envelope for a moment, recording the moment in his memory.

"Shall I send it?" he once again asked Anne.

"Is it the truth?" she said.

"Yes."

"Then send it."

He slipped the letter inside the box and it disappeared. His outrage was now final.

The Politician

Howard Bennett learned about Willard Bascom through the newspaper articles regarding David Brown's accusations that sprouted up for a few days and, later, through phone conversations with Brown himself. Bennett now had a villain for his Santa Monica Bay morality play. Unlike Mayor Tom Bradley or the Los Angeles City Council, who occupied a gray area between good and evil, the Bascom story was black and white. Whenever Bennett talked about Bascom, his voice got louder and the words sped by, expressing such umbrage that you would think Bascom had personally defiled the beach outside Bennett's house. Occasionally, he'd even feign spitting, which was as simple and shocking a gesture of disgust as he could come up with without using words.

Again, wanting not to be one who merely rants, Bennett sought what he saw as the perfect solution: he wanted, as he put it, to "prosecute this Bascom and make him swing in the wind as an example for other environmental criminals." So he, Carla Bard, and Ruth Lansford called at Tom Hayden's office at the Santa Monica Mall, where Bennett pleaded with the assemblyman to sponsor a bill making it a crime for public servants to lie in public. According to Bennett, Hayden didn't take the same angry delight in the idea of retribution visited upon Bascom, and said, "He's an old man. Just let him retire."

Bennett classified Hayden's reaction as politically motivated; that is, Hayden didn't see any career advantage to pushing such a law. (Hayden doesn't recall the meeting and now barely remembers Bennett: "Isn't that the guy who claimed he was swimming in shit?")

Just the same, most environmentalists at the time saw Hayden as the pivot man in advancing the idea that Santa Monica Bay had a better use than as a toxic dump site. First elected to the state assembly in 1982 representing the Santa Monica area, Hayden wrote several laws to protect the bay (which he now shrugs off as being somewhat ineffective) and unsuccessfully pressed the EPA to declare the bay a Superfund site. He held press conferences at the Santa Monica Pier. He pushed and he pushed, networking with the likes of Rim Fay, Dorothy Green, and other environmentalists, as well as the president of the Los Angeles Board of Public Works, Maureen Kindel, and, eventually, Mayor Tom Bradley.

"We had a lot of press conferences on the pier," Hayden says now. "I wore out my welcome in Santa Monica. They thought I was bringing a bad image to the city. But we were trying to build a movement."

In the early 1980s, however, he knew little about the bay's polluted state. He now tells a perhaps apocryphal but cinematic story of eating lunch at a Santa Monica seafood restaurant on

Former California state assemblyman and senator Tom Hayden, 2008.

the beach when he heard a radio news report of a huge sewage spill that went into Ballona Creek, which flows into the ocean near Playa del Rey beach. This, he says, was when he first discovered just how bad things were.

It gets a little murky from there, because he doesn't recall when this happened. The details he relates of a sewage overflow box popping its lid—by design, actually—first came to public attention in July 1985, two months after the May 17 task force hearing, where Brown and others told him details more frightening than an overflow box. (According to Green, Hayden and she held a joint press conference at the overflow box.)

Just the same, Hayden dipped his toes in the bay's dirty water more often than any other politician of the period. "You don't get elected around here without giving lip service to the bay," he says now, feeling a little cynical about what he considers a lack of progress over the years. He blames that on spineless politicians and bureaucrats, if not environmentalists, such as Green, who he believes have sold out to negotiate with the pollution perpetrators instead of just taking them to court.

"Lawsuits have this clarifying quality," he says now. "There can be plea bargaining, but only after an acknowledgment of guilt or responsibility. But there's no smoothing it over. You're either in violation of the Clean Water Act and state laws, or you're not." But when he held the task force hearings, the forty-five-year-old Hayden was still a little innocent or, perhaps more to the point, idealistic about his chances to clean up the bay. A likeable man (more so in private, when he wasn't caustically pitching his issues to the cameras) with thick, dark hair and a medium build, he was perhaps known better as actress Jane Fonda's husband (they divorced in 1990, after nineteen years of marriage).

And for those with longer, less starstruck memories, he was a member of the Chicago Seven, who were convicted of riot-making during the 1968 Republican convention in Chicago and then acquitted on appeal. In the California assembly, he was the iconoclast, often seen as so ultraliberal that he merely lurked about the fringes, barking earnestly about one social issue or another.

By 1985, though, with nearly three years in the assembly, he had managed to pass an eclectic array of legislative bills, some of them so unusual that it's possible no one else would have thought of them. Some of the thirty-one statutes that Hayden got on the books included programs designed for Vietnam veterans exposed to the toxin Agent Orange, and others concerned Neighborhood Watch training, an "Asbestos in the Schools Awareness Week," and record-keeping requirements for kosher fresh meat and poultry.

About a year before Hayden's epiphany at the seafood restaurant, the state legislature had passed his Santa Monica Bay Revitalization Act, which was his first attempt at dealing with the pollution issue. At the time, he told the *Los Angeles Times*, "It's very serious. What's of concern to me is that we would let a resource like this—a jewel in West L.A. and Santa Monica—run down with so little environmental and resource planning. I think the resource can be recovered, so I'm an optimist."

Well, that was 1984. He now concedes, "That was probably before I even understood the magnitude of the problem. I was just getting into it." It was indeed a small, almost passive, start. The legislation created a twelve-member Santa Monica Bay Advisory Committee that would make recommendations based on the state's Department of Fish and Game studies and input for restoring the bay and increasing the fishery.

Willard Bascom, perhaps feeling territorial about someone else snooping into the bay's aquatic health, grumbled to the *Los Angeles Times*, "If the premise is that we don't know enough about the bay, it's a mistake." In retrospect, too much knowledge wasn't the problem, and people such as Rim Fay and, later, Howard Bennett would accuse any number of government officials of a plot to keep the pollution secret. At the time, Hayden wasn't buying the paranoia. "It could also be that somehow we have not raised the issue," he said in early 1985, not understanding his statement's irony—the issue had been raised over and over without attracting that much interest. "I don't think it's true that there's a conspiracy in Washington [by EPA] to ignore this. I just think nobody knows about it."

Hayden's next step, the task force hearing, yielded more information than Bascom would have preferred. Of that Investigation of Toxic Pollution in Santa Monica Bay, Hayden's website claims, "These initial hearings revealed for the first time the systematic and massive pollution of Santa Monica Bay while state regulators did nothing." Narrowly defined, that was true. Ignoring that Fay, Bennett, and others had screamed about pollution on several occasions with varying degrees of detail and public attention, the hearings' most significant disclosure came minutes before it ended, when Brown nervously mumbled a few words about Bascom's threats.

Later, in his letter to Hayden, with more assurance, Brown accused Bascom of fiddling with the facts. Hayden received the letter the following Tuesday but resisted acting on its juicy contents until, according to Brown, he called to ask, "Dave, are you sure about this stuff?" In other words, Bascom's aura of credibility still hung about, and even Hayden, as cynical as he was about the establishment, as it was known then, wanted to make sure this oceanography icon had actually been as duplicitous as Brown said.

"I think he was afraid, you know," Brown says. "At one point, I think he didn't know whether he—this will tell you what the situation was—he didn't know if I was out there blowing my horn and it was a lot of hot air." Once Brown assured him he had carefully stuck to the facts, Hayden's office forwarded Brown's accusations to Robert Ghirelli, then executive officer of the Regional Water Quality Control Board, with a cover letter suggesting that the board and the task force jointly grill Bascom about the allegations. Later, Hayden would characterize Brown's allegations as having the "stench of scandal," and he advised Bascom: "I think for the sake of his agency's integrity, he [meaning Bascom] should step aside now." For his part, Bascom called Hayden an extreme environmentalist and refused to quit.

About a week later, a somewhat chastised Ghirelli told the *Times'* Alan Citron (as reported in "Are Bay Fish Safe to Eat? Showdown Expected Tuesday"), "There's going to be a concerted effort to move ahead. We really need more and better information on what's out there now. What are the [contamination] levels and what do those levels mean?" Ghirelli says now this prompted the Water Board to be leery about simply rubber-stamping the 301(h) waiver.

Brown expected Hayden to run with the letter's accusations, but naively figured the assemblyman would leave him out of it. But less than a half hour after the two spoke on the phone, the letter went to a few reporters. This included KCBS-TV's David Garcia, whose assistant immediately called Brown and, without mentioning they had seen the letter, asked for an interview to "just see what makes you tick." Brown and Garcia met later that day at Venice Pier, where Garcia admitted he had more than Brown's personal motivations on his mind; he had read the letter.

Reciting the story that night, the reporter couldn't resist a little opening self-promotion: "In recent months since Channel Two News first reported the contamination story . . ." (Near the end of the report, Garcia conceded that the public still didn't know what was going on, suggesting that stories about DDT still weren't making an impact.) For a somewhat contrived visual, a news camera recorded Garcia and Brown strolling down the beach, the scientist primly dressed in a short-sleeve blue shirt and tie. Brown, his thick dark blond hair blown back, said in response to the reporter's question about how he felt, "Just a sense of outrage that that information can be continued to be presented even though we know that a lot of it is only part of the truth."

The next day, on May 22, the *Los Angeles Times* took its own copy of the letter and printed Brown's charges without further comment from either Bascom or Brown, other than to report the obvious: "People familiar with [SCCWRP] say that relations between Bascom and Brown have been strained for several months." The next day, they tracked down Bascom in Vicksburg, Mississippi, who said, "I want the truth out, and I'm willing to take my licks in front of a jury of my peers. What I don't like are these wild allegations."

Bascom's peers—eight of them, at least—indeed met the following week to review Brown's "wild allegations." Unfortunately for Brown, Bascom chose the scientists on this so-called blue-ribbon panel himself, stacking the jury with his cronies. They included Professor Edward Goldberg of Scripps Institution of Oceanography and Professor Roger Revelle, the former Scripps director who had hired Bascom some thirty-five years before. For all the apparent importance of the blue-ribbon panel, they in fact had little authority to do anything but listen to the two sides and pass judgment on whether Brown's accusations had merit.

The day before this panel convened, the SCCWRP scientists furtively passed among themselves a brief statement supporting Brown's actions, which they were asked to sign. Fourteen senior scientists signed this "press release" and, according to one of the signers, Dr. Bruce Thompson, two abstained. In typical scientist fashion, the document parsed their sentiments carefully, with one blinding exception: "We regret the way that this controversy has been presented in the public arena. However, given the fact that it has occurred, we stand behind the data presented in Dr. Brown's letter to the Santa Monica Bay Task Force." The letter was dated May 28, 1985, and it was sent to the press.

Brown felt this was faint praise for his risking his career, especially given that about six months earlier the same group had declared they would rather lose their jobs than have their research withheld from the public. He felt even less love from some of the staff—such as lab technicians and administrative people—who mirrored Bascom's resentment over what he called Brown's publicity-seeking intentions, which in the science world was—and still is—considered extremely gauche. Those in Bascom's camp ostracized Brown, talking to him only if necessary.

Although years later Brown says he doesn't regret any of what happened, he was briefly stunned by the circus he entered that day. Bascom had decided not to hold the panel in a comfortable hotel conference room, with its less-than-comfortable expense, and instead cleared out part of a warehouse at SCCWRP's headquarters and set up cheap folding chairs on the bare concrete floor. As if to accentuate the big-top atmosphere, he had an orange-and-white-striped parachute hung from the high ceiling, presumably to make the room feel less cavernous or perhaps to hold down the echoes. The parachute fluttered in the breeze generated by a large industrial fan that stood at the side, substituting for air-conditioning.

The blue-ribbon panel—looking as august and scholarly as you might expect of gray-haired men who had spent their long lives immersed in science—sat around two folding tables with an audience of environmentalists, SCCWRP staff (who had left their posts to watch), and reporters sitting behind them. At times, Brown and Bascom both faced the panel, sitting within three feet of each other. They were caged by their professionalism, which required them to behave dispassionately, but their stiff body language shouted such discomfort and disdain for each other that it might have been more satisfying had they decided the issue with a fistfight. Instead, the two presented their cases over a four-hour period, accompanied by the multimedia of the day: slides and pre-PowerPoint graphs.

Brown opened his presentation with the same somewhat disingenuous reasoning he had been using all along to mask his rebelliousness. "I have not charged Mr. Bascom with anything," he said, reading from handwritten notes. "I gave the legislature task force additional information which I felt was essential, for them to have a complete understanding of the contamination situation in Southern California. I also answered questions asked by the task force regarding pressure. Let me add that I told the task force, if there was pressure, I ignored it." In other words, he was just a scientist trying to get the information out.

As the slide shows and graph presentations dragged on, the panel grilled Brown about his conclusions while mostly throwing softballs to Bascom. At one point, a panel member asked Brown if he had tried to quietly slip the information to various health agencies rather than publicly question his boss's integrity, implying Brown could have handled this more discreetly. "We have not found a means of effecting change," Brown replied vaguely after touching on the staff's frustration over Bascom's revisions.

In contrast, Bascom heard praise for his twelve years at SCCWRP. He had "been pretty straightforward," as one member put it. For his part, Bascom stuck to his story that Santa Monica Bay, despite the daily sludge discharges and layer of toxic chemicals, was getting better. "I have said the coastal waters are in pretty good condition. . . . But the press comes out and says I have said there is no problem."

That slightly obfuscated the issue, given that Brown hadn't disputed the cherry-picked facts Bascom had dribbled out, but how he had put a positive spin on even the most damning evidence. Publicly, neither man disagreed with the data SCCWRP had collected, and, out of discretion, conceded this was merely a difference of interpretation. In fact, Brown believed without saying so that Bascom had skimmed off a few facts and molded them to support outright lies. It was that public duplicity that

Brown had come to despise. As long as Los Angeles got its 301(h) waiver, it didn't seem to matter to Bascom how distorted his statements were to the Bradley administration, the County of Los Angeles, or the public.

Strangely, even though the chairman of SCCWRP's consulting board, Dr. Perry McCarty, had arranged the hearing (with Bascom's input), no one really said what authority the panel wielded in the matter, other than, presumably, to decide if Brown had a case. "I don't even know what their purpose was," Brown told reporter Larry Keller later that day. Just the same, press coverage was heavy, with sound bites showing Bascom calmly repeating his routine assurances, a clunky microphone stand against his chest, and Brown appearing a little worn as he answered reporters' questions.

After the presentations, the panel went into a private session while the audience milled about the room. During their wait, a SCCWRP administrative assistant found Brown and sternly asked him, "Do you really know what you're doing?" Brown, by this point feeling defiant, and perhaps annoyed, stared at her and snapped back, "I know *exactly* what I'm doing." She stormed off.

"At that point, I did know what I was doing," Brown says now. "I knew exactly what I was doing, and I knew that the story had to get out and I was going to get the story out. . . . It wasn't, like, accidentally that I got the story out. I got the story out exactly as I intended."

The panel couldn't decide if they liked that story or not. They came back about an hour later and said they hadn't made a decision yet on whose version they were buying. As they put it, five members agreed on an as-yet-undisclosed consensus and the other three weren't so sure.

The next day, Brown went back to work, a pariah to some, a hero of sorts to others. It's not that he expected to be congratulated, but he hadn't realized people on the SCCWRP staff would actually shun him. As time went by, fewer and fewer even bothered to say hello. Nevertheless, he remains convinced he did the right thing.

"I was angry," he says now, "because I realized my research, which was really to look at mechanisms of detoxification and toxicity, was being used by Willard and the supporters as a panacea for the discharge of contaminants into the environment. So I was being used. My research was being used, and I was angry about that. But I was more angry about the fact that Willard— I'd use the word—lied to the public. There is no other word to use. I could clean it up and say, like I said on camera, it's all part of the truth, but nah, he lied. And I was really angry about that. That we would be doing this research, and we had these facts, [and] nobody in Southern California knew what was going on in this massive oceanic environment that was so contaminated and so overwhelmed with biologic material from human sources that it was getting highly degraded. That made me angry."

John Dorsey was disappointed, but for different reasons. "I was watching all that from the sidelines, of course," he says now. "Personally, I was kind of sad all that was happening. I thought this is really sad because SCCWRP's a great organization. And to have all this going on just sucks. Not a good deal. Because it could very well drag the science down, and it's going to make people suspect the science. And I was really afraid of that because they did really good work. Because people were saying, ah, they're being paid for by the dischargers. You can't trust them. Yeah, they were being paid by the dischargers, but they were all really good marine biologists doing very honest work. So I was very sorry, sad, that episode happened like that."

Two days later, no longer in front of cameras, the blue-ribbon panel released a statement saying, "The panel found no evidence Mr. Bascom deliberately misled anyone or knowingly supplied false information or withheld information that should have been published. It is common for wide differences of opinion to exist when environmental data is interpreted by knowledgeable scientists of good will and with good intentions. We believe the differences that have surfaced recently are of this kind." And they then took a shot at Brown for his behavior. "The panel believes the procedure used by Dr. David Brown to air his concerns over the scientific decisions and practices of SCCWRP and the integrity of the Director is unfortunate."

Publicly, SCCWRP's board announced that both Brown and Bascom had suggested a thirty-day cooling-off period. Brown says, in fact, this was a forced vacation, and that when he objected, someone associated with the panel let him know that if he didn't accept the implied penalty he would be fired instead. As a result, when his two sons later came down to visit for the summer, he wasn't able to see them except before or after work and on the weekends, and for him this was the worst consequence of the letter he wrote Hayden.

Meanwhile, Bascom was essentially told to resign immediately, which had little consequence, given that he was retiring less than two months later, in July. More to the point, Bascom hadn't yet attained his ultimate goal of securing the 301(h) waiver for Los Angeles. SCCWRP's jury of his peers may have absolved him, but Brown's letter and all its carefully worded accusations had more staying power. Howard Bennett believed his opposition to the waiver had been indirectly blessed by the letter, and this gave him much more ammunition to fire at the city's 301(h) waiver application. Environmentalists saw Brown as their champion, the necessary inside guy blowing the whistle on his bosses. The public started to eye the beaches with doubt, wondering if some disease-causing microbe had their name on it.

But no one knew what the Regional Water Quality Control Board thought and whether they would still approve the waiver.

CHAPTER 11

The Brown Ribbon

It should have been just a lousy phone bill. One of those normal obligations that most people pay without thinking twice. But this—this was a vacuum hose inserting itself into Howard Bennett's checking account and sucking out the dollars. Pages and pages listed long-distance calls that he had sprayed like buckshot to Southern California, Sacramento, and Washington, D.C. Government officials, legislators, environmentalists, and the media were all represented by cold black numbers and indecipherable city abbreviations. The total due amounted to nearly what a schoolteacher took home in a month, and it was easily the single most tangible piece of evidence of just how much Bennett had put into his campaign.

Sure, he spent hours cajoling people to participate in the campaign and imploring the press to open their jaded eyes to the issue in which he believed so fervently. But measuring the time he had put into something this important seemed almost petty and hardly relevant; he could only describe the cost, grandly, as a necessary sacrifice. His days and nights blended together into a life-consuming intensity that he didn't bother to quantify.

But a phone bill was like an audit on his commitment. Spending more than three thousand dollars in phone charges and other expenses left such a blister on Bennett's memory that he brought it up years afterward, as if, for storytelling purposes, a wad of cash represented his efforts in one tidy image. Yes, he could describe talking on the phone for hours, meeting with environmentalists for hours, or getting up at 2 A.M. to write down a publicity idea. But the feeling one gets when three thousand dollars disappears from the bank account—that means something. It's a tangible pain. Measured by money, his labor meant something.

Bennett considered asking for donations. "We need money," he scribbled in a note to himself, drafting an appeal that included all the ways he personally contributed to the campaign. He wondered if perhaps he could hold a lottery to raise the cash. In the end, he decided not to put his hat out for the few quarters he might get. He saw himself as fighting an injustice, and in that lofty context the only compensation one sought was the accomplishment of one's goals.

In that respect, at least, he could point to real progress. He had achieved his first big objective—to force a second public hearing on the 301(h) waiver—so spectacularly that it had become the media event he had desperately wanted. Moreover, the Los Angeles Regional Water Quality Control Board had heard in no uncertain terms just how he and the rest of those there felt about Hyperion's sewage treatment.

Unfortunately, the board said it wouldn't decide on the waiver before July. Being a suspicious guy, Bennett figured that, if the publicity wilted away after the second hearing, the city, state, and EPA might still sneak the waiver through as though the protests had been a minor blip in public opinion. If everyone was distracted by some other issue, who would do the follow-up? Like a salesman trying to close the sale, Bennett figured he had to keep applying pressure. But in order to truly put the screws to the decision makers, that pressure had to be public. It had to be big. It required press coverage.

He gathered the two people he trusted most, Bente and Leif, at the small, Scandinavian-style table that separated the kitchen from the living room. And he presented his idea: a five-mile brown ribbon stretching out from city hall toward the ocean. It wasn't just a ribbon, he pointed out, so excited now that his hands couldn't stop gesturing. It had to be covered with petitions imploring the mayor and city council to spend the money, to rebuild Hyperion into a full secondary treatment plant. Think of the symbolism! The ribbon represented the five-mile discharge pipe from Hyperion out into the bay, where it currently poisoned the fish.

It was outlandish. It was visual. It was video friendly.

It was also about four miles too long.

Bente and Leif looked at each other before giving their simultaneous critique: "Perhaps you should scale down the ribbon's length. Five miles? That's going to cross streets, stop traffic. People will lose your message if you inconvenience them." The two then squeezed Howie's enthusiasm into a small bottle labeled Reality and provided the practical analysis he seemed incapable of doing himself. The ribbon was a great idea, but, one of them suggested, he could wrap it around city hall instead.

Now that's symbolism! Think of the impact! Think of the fact it doesn't need to be so long!

Bente and Leif were perhaps the only people who could shake loose Bennett's ideas from his blind enthusiasm and remold them to fit the real world. So once they convinced him the five-mile ribbon was too long, Bennett cheerfully figured a one-mile ribbon could represent the infamous one-mile overflow pipe instead. Even better. (The one-mile pipe dumped chemically treated sewage in the bay when volume at the sewage plant exceeded its capacity, or raw sewage on rarer occasions.) The three then plotted out the demonstration as if it were a field trip, putting together a loose plan for assembling materials, gathering participants, and collecting petitions.

Bennett and his son then drove to city hall, and Leif calculated the ribbon's exact length. "Hey," Bennett would say later, "if you have a physics major in the family, you might as well take advantage of his math skills." With an eye on the odometer of Bennett's Chevy Impala, they drove down Spring Street in front of city hall and then turned right on Temple, at the building's side. The two blocks measured about half a mile. Leif doubled that to include the other two sides, and announced they needed a mile-long ribbon. Such mathematical prowess impressed his proud, numerically challenged father.

From here, the details piled up. They had to find a mile-long ribbon strong enough to hold petitions stapled to it and not tear or break as people stretched it around city hall. They also needed a wheel of some sort on which to roll up the ribbon and petitions, so the package could be safely transported to the protest and then unrolled. They had to let the police know a crowd would be showing up. Given that this was a large-scale project, Bennett figured they couldn't get by with just a few petitions. The petitions had to be so numerous that they would hang along the entire length of the ribbon like voters nipping at the mayor.

Bennett once again turned to his students for help. He announced a schoolwide contest: the person who got the most petition signatures would win a trip to Hawaii. Once again, his money funded the campaign. He called on his tenant, a teacher who happened to moonlight as a travel agent, and asked for the best deal out there for a week on Oahu. Another several hundred dollars disappeared from Bennett's checking account, and for that price he collected more than five thousand signatures, which he sent off to the Regional Water Quality Control Board's chairman, James Grossman. Bennett included a cover letter that, among other things, attacked Willard Bascom: "[He] has attempted a whitewash that won't wash. . . . Public opinion demands that you disregard the distortions of Willard Bascom!"

Dorothy Green's husband, Jack, constructed a wooden spindle for the ribbon and also bought the eight-inch-wide satin ribbon. Since it came only in thirty-foot lengths, Green walked away with 176 rolls. He found a brown-colored ribbon ugly enough to not only represent a sewage pipe but bring to mind the color of effluent as well. Bente, who owned a sewing machine, triple-stitched the reels and reels of ribbon together. The original petitions were photocopied before they went to Grossman, and volunteer students helped staple the copies to the ribbon.

Finally, one late Sunday morning, July 2, about 175 demonstrators, many associated with the high school—students, teachers, and parents—met near city hall's sunny front steps. Bennett arrived, his eyes behind his thick, wide glasses excited and anxious. "Let's get this going," he said out loud. "Are you sure it's going to work right?" he asked Leif. "I wonder if the mayor will show up. Probably not. I know what to do if he doesn't."

The crowd coalesced around the hyper Bennett as if they were all his students. Wearing a light-brown sport coat, white shirt, and tie, he reviewed the protest's goals, what the ribbon represented, and how they hoped to get Mayor Bradley's attention, assuming he was anywhere near city hall on a Sunday afternoon. Yelling out orders, Bennett exuded authority and presence, the alpha male leading the herd. People respected him. They listened to him. They drew upon his enthusiasm for their own.

Bennett then introduced Leif as the man responsible for figuring out the ribbon's logistics. Knowing just how simple that was, an embarrassed Leif instructed the crowd on how to unfurl the ribbon so that it didn't sag or touch the ground. After all, this was an aesthetic thing, really, the visual of a "sewer pipe" around city hall. One last detail had taken Leif and several friends from Caltech a couple of hours to calculate: just how much torque was needed so no one pulled the ribbon too taut or too loosely as they carried it down the sidewalk.

So he told the crowd that each one had to watch the person in front of them and make sure the ribbon between them was relatively level. Don't worry about what's going on behind you, he said. Separate yourselves by about thirty feet. With that, Leif, dressed in a bright yellow sweatshirt and black pants—a photogenic combination, he figured—led the procession down the sidewalk.

The brown ribbon protest, 1985. An unidentified protester helps stretch the mile-long ribbon around Los Angeles City Hall.
Photo courtesy of Bente Bennett.

At first, he and his fellow geeks looked like the geniuses everyone assumed they were. For the first quarter mile, the ribbon flowed along at just the right speed, holding level and taut. Some people even multitasked by carrying signs reading, "DON'T FLUSH IN MY OCEAN," "NO MORE SLUDGE BAY," and, on the bad-boy placard, "DON'T S—T IN OUR SURF."

But then, Leif noticed, the ribbon started to pull him backward. He and three friends in the lead tugged a little harder to keep from getting yanked onto their butts. They hadn't taken into account the accumulated pull that dozens of people behind them would exact on the ribbon. The strain on the ribbon increased as they continued around the block. Somehow, Bente's sewing job held.

While this was happening, King Neptune, god of the sea, arrived. Patrick Wall, another man who understood the value of a good visual, dressed up in a brown robe he borrowed from his wife, yellow socks and flip-flops, and a ratty white beard and wig that, in another context, might have been cheesy but here somehow fit in. He carried a plastic pitchfork and sign saying, "GET SERIOUS ABOUT SEWAGE—NEPTUNE." As he strode down the sidewalk, he announced to the crowd, "Save my ocean!"

Bennett politely greeted the surprise interloper, who whispered to him, "Howard, it's me. Patrick!" But this wasn't part of the plan, and even though Wall thought he was making a clear statement, Bennett wasn't so sure the crowd would understand the message. Or perhaps it was just a little rude to crash another man's party and, for a moment, steal his thunder. Either way, King Neptune was absorbed into the proceedings, and the protest continued its slow march around the block.

Once the brown ribbon surrounded city hall, Bennett theatrically announced at the building's doorsteps that he had a stack of petitions to present to Mayor Bradley, who predictably didn't show. But Bennett already had a plan B. Positioning himself where the few cameras in attendance could get the best shots, he bent over a manhole cover—presumably one above a sewer line—and used it as a proxy for the mayor. Once the manhole cover had received the petitions, the demonstration was pretty much over, except for interviews with the reporters, whom Bennett welcomed as though he were a party host.

"There is no reason in the world why [the city council] can't do full secondary treatment," he bellowed. "They just want to save money at the expense of other people. It is an ecological disaster." Without much concern for a possible slander lawsuit, he referred to Willard Bascom as "a silver-tongued con man. He has lied to the city council and to the people."

Talking to KABC-TV's reporter, Bennett was a little more circumspect but just as acid-tongued: "LA has wound three pipes around its citizens. We feel it's only fair—turn-about fair play—to wind one around the city hall." However, as visual and caustic as the protest was, the station devoted only a scant 43 seconds to the event, with anchor Harold Green merely providing a voice-over to the video (as opposed to a so-called package, where a field reporter covers the story with both on-camera and off-camera narrative; the latter usually signifies that a program's producers consider a story important). But that was the thing about seeking media coverage. Even if the report had been longer, it still would have been transitory. It was a moment on a Sunday evening seen by a few people, with no guarantee they would remember it.

Just the same, years later, the people involved still reminisce about the ribbon around city hall, calling it one of the Bennett campaign's most creative gestures. It not only attracted brief media coverage but, just as important, also further stoked the environmental community's interest in the issue. Not that it had much impact on the Bradley administration. According to some on his staff, the demonstration went largely unnoticed.

The next day, as if to refute the demonstration, Willard Bascom published an opinion piece in the *Los Angeles Times* titled "Santa Monica Bay on the Mend," in which he tried to distinguish between the bay being contaminated (not so bad; technically speaking, everything is *contaminated*, nothing is pure) and the widely held belief it was "polluted" (not the case, he wrote). "This does not mean that conditions in Santa Monica Bay are as good as we might wish. But they are pretty good, and steadily getting better," his article said. "If no action is taken, natural processes will have it almost back to its original condition by the time any new treatment plant could be put into operation." Given that effectively Bascom no longer worked for SCCWRP, it was almost as if he wanted people to know that he hadn't distorted the truth for the dischargers' benefit; he really believed this.

The next week, the *Times* published Tom Hayden's reply, in which he compared Bascom to the deranged movie character Dr. Strangelove: "Bascom's strange love is of toxic-laden sewage, which he believes provides nutrients that allow marine life to multiply and prosper when dumped into the ocean."

About the time this debate took place, Los Angeles missed its court-ordered deadline to stop discharging sewage sludge into Santa Monica Bay. The EPA, rather than fine the city a thousand dollars a day, negotiated a new deadline, for February 15, 1986, seven and a half months away. The city blamed construction delays at Hyperion, where they were building HERS, the EPA-funded gizmo to dry and burn the sludge.

The *Times* called Bennett and, seemingly to make up for the fact that it gave the ribbon demonstration a mere inch or two of column space, asked for a comment on the missed deadline, in a sense crowning him prince of reliable quotes. He told them with his usual flourish, "To continue to allow Los Angeles to dump sludge into the ocean . . . is another example of callous disregard for the public. What's happening now has happened for years and years. It's 'tomorrow, tomorrow, tomorrow'—and tomorrow never comes."

The coalition he started now had credentials with the press. But, while the movement itself had a kind of self-sustaining energy, Bennett was exhausted. He didn't know how to have a conversation anymore without talking about Hyperion, the pollution, or the politicians. He needed a rest. He was about to hand off the coalition to Dorothy Green, and his days of confrontation were almost over.

CHAPTER 12

Heal the Bay

Bente Bennett felt as though she had lost her husband. He rarely talked about anything other than the campaign, and for months life had centered on sewage. Sewage! Yes, she was angry that the government in all its forms was responsible for her Howie swimming in polluted waters. But this obsession with sewage had strangled their life, their marriage.

It wasn't just their relationship. It was the constant intrusive phone calls. Howie had scattered calls seemingly across the country like a telemarketer, and now people were calling him back. And even though she disliked talking on the phone, she felt obligated to answer. She had become Howie's secretary, writing down messages or listening to people yammer on about the pollution or how awful it was, instead of just simply leaving their names and numbers, *and then hanging up!*

Every dawn, Howie still ran off across the cold sand for his daily swim, and she worried about all the one-celled horrors sliding over his body looking for a cut or convenient orifice in which to deposit themselves. Despite the talk of surfers and life-guards getting sick from waterborne bacteria, he stayed healthy, and she wasn't sure if that was a testament to Howie's unusual sturdiness for a guy fifty-five years old with the stress levels of a combat soldier. And that stress should have killed him by now. After all, Howie didn't know how to focus on something and still maintain a balanced life. Leif later called him monomania-cal. Yeah, that about described it. As soon as she had encouraged her husband to take on this campaign, she had lost him. So, after the brown ribbon rally, she wanted him back. All of him.

It started out as scattered conversations that built from com-ment to comment, her fractured grievances slowly expressed over time. Howie may have heard her remarks about his obses-sion and how it bothered her, but they had nothing to do with the campaign, and so her frustration went unheeded. Finally, she gave him that Danish death stare of hers, with the mouth pursed, the eyes squeezed down, and the anger about to shoot out in thunderbolts. It was time for him to throttle down, to pay more attention to her and their marriage. He had won the sec-ond 301(h) waiver hearing, and now he needed to spend some time thinking about her instead.

They sat at the kitchen table, where she told him it was for his own good as well. This was the kind of choking fixation that could destroy his health. He needed a break. *She* needed a break. Implied in all this was that their marriage—despite the passion-ate love they felt for each other—would become shaky if it con-tinued to be pushed aside for the sake of a campaign over pollu-tion. Forget about Santa Monica Bay for a moment, she pleaded. Hyperion was not to enter the conversation. *We're talking about us, about your obsession, about walking away.*

Bente also knew Howie wasn't capable of simply taking a break from the campaign. He would have to actually give it up and clear it all from his mind. She strongly laid out her feelings with the kind of pragmatic urgency that Howie lacked. She knew, too, that unless they left town, he wouldn't be able to distance himself from the campaign. For perhaps the first time in weeks, they put together a plan that didn't involve sewage. They pulled out an atlas and charted a flexible travel itinerary, planning to go where no obligations existed, on a voyage that didn't require a schedule. No phones. No crowds.

For his part, Howie knew she was right. This was no life for Bente, he told himself later. She had suffered enough. Still, he couldn't abandon the campaign entirely. The issue had to continue getting publicity. The city had to be reminded that its mayor and city council were neglecting the bay's environmental health. His anger had to continue through someone else.

He first called his neighbor Ruth Lansford and asked her to take over the Coalition to Stop Dumping Sewage into the Ocean. Given her full-time involvement in fighting development with Friends of Ballona Wetlands, she declined. Like a salesman needing that last commission, Bennett immediately went down his list, calling others who shared his indignation, but they all turned him down for pretty much the same reason as Lansford—they were too busy. This was a common affliction among activists. They had to choose just one issue and make it their own, or the fight in them was diluted. And while none of the people Bennett contacted dismissed the bay's pollution as unimportant, they had their own obsessions. In some cases, they also had spouses or families who were perhaps a little tired of one idealistic cause after another, and would threaten divorce if anything else were now to claim the few moments in the day they had to simply, freely breathe.

According to Bennett, Dorothy Green's name sat low on the list, as though he didn't see her fitting into the campaign as well as some of the others. In a way, that was true. The two had become more and more estranged as she and her League of Conservation Voters board argued with him over which direction to take the campaign. By this point, they had tired of his insistence that they follow his lead. According to Green, she and her husband, Jack, took Howard and Bente to dinner at a Marina del Rey restaurant, and "we spent the whole evening talking about what it means to be a coalition and how to work cooperatively and to do it. . . . And he told me he was incapable of working that way in a coalition."

Bennett stared at her name for a moment. The coalition was his child, and it felt as though he was about to abandon it at the doorstep of a woman he was unsure could care for the campaign with the same love he had given it. Still, he needed a proxy for the fight, and Green—along with her organization's board— had plunged into the issue with probably more enthusiasm than anyone else. He paused a moment longer and then called Green. With the kind of sales patter he had learned years before, he introduced the deal to her as though she would be foolish to turn it down.

For Green, this meant an opportunity to fold the coalition into her league's previous efforts to kill the 301(h) waiver. Bennett saw it a little differently: she was simply taking over his organization and would carry on the same loud, theatrical tactics he had employed thus far. It was perhaps a fine distinction, but this came down to ownership, ego, and, ultimately, control. Green didn't deny Bennett had started it all, but she saw him as quitting the cause just as the momentum needed for the waiver's defeat had built to an angry, well-publicized peak. By taking over the coalition, she now had a chance to divest the campaign of Bennett's confrontational style and endow it with what she considered a greater sensibility. For his part, Bennett assumed the coalition would continue to use his methods. As far as he was concerned, it didn't need any fine-tuning.

Neither of them mentioned their expectations to the other. Instead, without the hesitation Bennett had felt in dialing her number, Green accepted the offer. She officially took over the Coalition to Stop Dumping Sewage into the Ocean.

At first, this didn't mean much. The coalition merely became a subset of the league, with Green and her cohorts meeting periodically to discuss different approaches to tackling the water pollution issue. Not much changed there; they had already been doing that for the past two months. The coalition member organizations—who were there on the letterhead mainly for the appearance of strength in numbers—were no more involved with Green than they had been with Bennett.

Believing the campaign would continue at full strength, Howie and Bente secured two open-ended plane tickets that allowed them to fly wherever they wanted, when they wanted, for six weeks. They started in Iceland, then went to Greenland, the Faroe Islands, Scotland, England, Portugal, and Spain, and finally they went snorkeling in the Grand Cayman Islands.

In a letter published in the Sierra Club's Clean Coastal Waters Task Force newsletter, Bennett told his supporters, "I will be leaving the country for the summer and won't be back until September 8th. Dorothy Green, President of the League of Conservation Voters, has graciously consented to be the Acting chair of the Coalition during my absence." While calling Green the "acting chair" made it appear as though Bennett intended to take back the coalition once he and Bente returned, Bennett says that, at this point, this was his last farewell, he had left the fight for good. But that would not be the case.

Meanwhile, a small slice of the activists from the League of Conservation Voters board continued to meet in Green's luxurious living room, slipping and sliding over the kinds of organizational issues Bennett never worried about, or had to. And indeed, this was Green's forte: to take an issue that was an object of simple mob protest and turn it into something with card-carrying officers and members, and then produce a kind of self-sustaining impetus among those disciples and a common drive to meet certain spelled-out goals.

"Dorothy has a tendency to rule," says Moe Stavnezer, who attended many of these first meetings. "And I'm not saying that in a critical way. That's who Dorothy is. She's a leader. She's imaginative. She deserves any accolades that you can possibly give."

With Green in charge, the board fretted over whether or not the coalition really fit with the league's focus on supporting green politicians instead of involving itself with actual environmental issues. In the kind of discussions that would have bored Bennett, they tried to decide how to break away from the league and form their own group, apparently not giving much thought to simply reformulating the league's goals.

After Howie and Bente returned from their trip, the league put out an announcement dated September 1985 that read in part:

> Because of the work or *[sic]* one man, Howard Bennett (a long distance ocean swimmer who lives at Playa del Rey) a second hearing was granted. . . . His efforts are paying off in several ways. A new staff report has been ordered that will assess all the new information presented at this hearing for EPA and the Water Board to use in their decision making process, and more and more people are getting involved. One dedicated and angry citizen can mobilize and energize people to fight City Hall. There is now a very active and growing group of organizations and concerned citizens who are organizing to defeat the waiver, and to restore Santa Monica Bay so that it will once again be safe to swim and the fish good to eat. This coalition has formed under the auspices of the Los Angeles League of Conservation Voters.

Eventually, this small group approached the wordy name Bennett gave the coalition (in the above press release, they didn't address it by the full name). The small group no longer wanted to spit out "Coalition to Stop Dumping Sewage into the Ocean" every time they brought up the subject, and felt they needed a moniker that was easily remembered and, well, didn't include the word *sewage*. The name also required a publicity-friendly quality that stated their mission in a way that was memorable yet not too confrontational. Someone suggested they call themselves Save the Bay, which copied a San Francisco group's name. A clunky lifesaver logo came with the quickly killed idea. In fact, they nixed every name they considered.

This included someone's somewhat lame proposal, "Heal the Bay." No one said anything positive about it, other than that it was the least objectionable candidate, and to this day no one in the group can remember who thought of it. In a time when the self-involved, easily mocked New Age was running its course, the idea of healing the bay seemed almost like a joke. It sounded as if they might try cleaning up the ocean with crystals.

And yet the idea kept coming back to them. It wasn't that it grew on them or someone championed the name to such a point that they were all infected with the same enthusiasm. It was still the same old feeble *Heal the Bay*, but nothing sounded better. Finally, feeling as though they had better things to do, like fighting the 301(h) waiver, the group collectively shrugged and agreed to call themselves Heal the Bay. Hopefully, no one would laugh. The moniker would later come to represent the entire environmental movement to many people—particularly in California—but no one in the group takes credit for the idea or can remember who thought of it. One of the league's many vice presidents at the time, Jamie Simons, says vaguely, "I think a couple of people kind of weighed in on the name." A reporter for City News Service, Marc Haefele, who had been following the waiver story from Bennett's first press conference, wonders aloud if he might have inspired the name: "You know, I have this weird idea that I can't substantiate, that maybe I thought of that name. I just remember thinking, Wouldn't it be nicer if they could say 'Heal the Bay' instead of the 'Coalition to'—and it's possible I told somebody. I don't know if I can take credit for it."

However the name came about, Simons says it reflected Green's less confrontational style compared with that of Bennett. "I do remember Dorothy, always, in anything she approached, felt like people had to come together. Things couldn't be antagonistic. You had to sit down with business, and it couldn't be another environmental, we-hate-you, you're-the-bad-guys. It had to be everybody really talking. And everybody really trying to understand each other's point of view. And so, I think, out of that idea [it] really became, like—it kind of had a bigger meaning."

Next, Bennett had to be told. Green called him over to her house, where the rest of the nascent organization met, and they revealed the coalition's new name. He cringed. He choked. He hated it. Not terribly aware there was a New Age to begin with, he didn't react to the healing notion, but responded with his conviction that his former name worked in its all-encompassing nature. By calling themselves Heal the Bay, they had automatically narrowed their job down to just Santa Monica Bay. What about the rest of the world? He called the coalition "Stop Dumping Sewage into the Ocean" for a reason. It grandly meant any ocean, not just one spot on the map.

"Heal the Bay—is this the only bay that needs to be taken care of?" Bennett says now. "It's the whole freakin' west coast of America and the east coast of America and every city on every ocean in the whole world. Why miniaturize it or, if you will, vitiate it, or—let's get another word—cut it down by [demanding that we] just Heal the Bay? Santa Monica Bay—it is true the EPA called it the most polluted body of water on earth. Okay, but hey, it's more than Heal the Bay. But [Green] was fixated on that title. By that time, I had resigned, and Bente was very wise—[she] said, Keep your mouth shut and walk away." And so he did, but for years afterward, he complained Green had tossed aside his coalition's broadest goal.

On October 28, the league sent out this memo:

To: The Coalition to Stop Dumping Sewage into the Ocean

RE: Heal the Bay

With the decision on the 301(h) waiver now set for November 25 [it was postponed from July] that will determine if Los Angeles will be excused from providing full secondary treatment for all of its sewage, it is clear to many of us that the work begun under the leadership of Howard Bennett needs to continue beyond this decision date. The problems that afflict Santa Monica Bay will not disappear, and many of those problems are not addressed by the waiver process.

As the bay continues to become even more polluted, posing an imminent public health threat to all who swim or surf or eat fish taken from it, and threatening property values in the beach communities, we feel a growing obligation to educate the public and to increase the political pressure on those elected officials who can take the actions necessary so that the bay can begin the process of healing itself.

HEAL THE BAY GOALS

Therefore we are asking all the organizations who oppose the 301(h) waiver to join us in an expanded coalition we are calling "Heal the Bay," dedicated to achieve *[sic]* swimmable, fishable coastal waters that meet the goals of the Clean Water Act "to restore and maintain the chemical, physical, and biological integrity of the nation's waters."

TO ACHIEVE THESE GOALS, WE NEED

- A political action program to pressure Mayor Bradley, the City Council, Congress and our state legislature.
- An educational program to reach out to the community.
- A Membership program for both organizations and individuals that is inexpensive and therefore will attract a lot of people.
- Fund raising to cover the costs of our program.

In some ways, the group's goals were no less ambitious than Bennett's original plan. They also strongly reflected Green's contention that confrontational tactics were hurtful to the cause because they discouraged decision makers from noodling with Heal the Bay or others over how to jointly solve the pollution problems. In a way, however, Green and Bennett agreed on the education part, so much a part of Heal the Bay's goals. It was just that Green believed they could do something more substantial than tie ribbons around city hall. Single demonstrations mattered for only a short time. Indoctrinating the public with a concern for the environment would last for years.

As Stavnezer puts it now, "Dorothy had a queenlike aura about her and, to some degree, still does. Dorothy is really very respected by almost everyone. Dorothy is a control freak. So am I. But it has been a real big plus, because when Dorothy got involved with something, Dorothy *got involved* with something. This was not someone who did half-baked stuff at all. And I can't downplay [how] Dorothy's role in Heal the Bay was pivotal. There's no question. Just [using] her name gave us a recognition that, if it had [been only] me and Felicia [Marcus, another member], [the group] probably would not have gotten!"

He laughs and continues. "Dorothy gave us legitimacy. She gave us a name. She gave us expertise. And she gave us strong, consistent leadership."

Despite Green's implication in the group's first two communiqués that Bennett was more or less history, he still had one more demonstration left in him under the coalition's former chunky name, and that involved a toilet. Perhaps, to a showman like Bennett, using a commode to make his point about sewage was inevitable. But the Dirty Toilet Awards would make permanent any rift between him and Dorothy Green, a split so wide that, in the future, his name would never come up in Heal the Bay's official history.

CHAPTER 13

The Dirty Toilet Awards

Dorothy Green generally came across with a near-grandmotherly warmth so immediate that you half expected her to pull a plate of warm biscuits from her purse. Jamie Simons, among others, thought of Green as her "second mother." In some cases, this might have been mere strategy, but it did come from a learned conviction that, if you wanted to sway the decision makers, at the very least you had to behave as though you respected them. That's why Howard Bennett's plan to flush their names down a john like used toilet paper had to be stopped. This was more than just juvenile; it was counterproductive.

One other thing about the Dirty Toilet Awards: In her mind, they plunged so deeply into poor taste that she spat out her opinion of the affair as though a gnat had flown into her mouth. The awards were appalling, scandalous. You would think that Bennett had taken a bucket of malodorous sludge from the ocean floor and dumped it on Green's Louis XV chair.

But when she got beyond the taste factor, Green allowed that Bennett had the right to do this except for one thing: he still claimed the coalition as his own and continued to hand out pronouncements with twenty-seven coalition members listed on the letterhead, including the Los Angeles League of Conservation Voters. This implied that the league supported such sophomoric shenanigans, and the nascent Heal the Bay, still loosely connected with the league, might be besmirched by any negative reaction. If the Bradley administration perceived Heal the Bay as being behind the Dirty Toilet Awards, it might never give the group a seat at the table.

It was entirely possible that Green overreacted. Others found something amusing, in a low-class way, about dumping crude effigies of Bradley and the city council down a toilet—Bennett emphasized it was a *clean* toilet—and sending them a Dirty Toilet Award, suitable for framing or wiping . . . well, never mind. It was a visuals-galore kind of stunt, perfect for the cameras, which meant, of course, the press would report on his puerile theatrics with enough gusto that the public might be reminded that the EPA could still grant Los Angeles a 301(h) waiver. With perhaps a little too much bravado, Bennett also saw it as shaming the guilty characters involved into finally popping loose a few million bucks toward fixing the Hyperion Sewage Treatment Plant.

The story from here is blurred by differing, imperfect memories. Bennett remembers that he and Bente invited Green and her husband, Jack, to dinner, where he announced the awards to the aghast Dorothy (Jack seems to have said little). The awards, he says now, were sparked in part by his belief that Green had done nothing publicly to advance the campaign, which is true.

The Heal the Bay committee had spent their meetings strategizing, developing goals, and naming themselves, without staging any television-friendly demonstrations. In any event, Green tried to convince Bennett what a horrible idea the awards were, and when he insisted on going ahead anyway, she told him, "I will have nothing more to do with you."

In Green's slightly more dramatic version, Bennett blind sided her by inviting her to the awards ceremony with no warning as to its contents, and she stood in the back of the room shocked when he started sending the effigies into the bowl. For the record, when confronted with Green's version, Bennett emphatically declared it untrue. But both of them had nothing other than their memories to back up their stories.

Either way, on November 5 the awards went on as planned. Although he might have held the ceremony in a local Chevron's men's restroom just for yucks, Bennett once again rented a room at the Los Angeles Press Club. He borrowed an off-the-shelf, shiny white toilet from a friend who ran Snyder Diamond Plumbing, with the promise that he would return it in one piece, unused. The toilet simply sat on the dais, a conspicuous prop. The award certificates, which hung on a corkboard behind the toilet, were a bit crude, with a frilly clip-art border surrounding an open toilet and the inscription, in florid type: "Roll of Dishonor. This Dirty Toilet Bowl Award is given to [blank] because you have failed to withdraw the 301-H Waiver Application for Hyperion."

Besides the press, Bennett invited various dignitaries, who perhaps wisely declined the invitation. Michael Antonovich, Fifth District representative for the Los Angeles Board of Supervisors, replied in his RSVP with an apparently straight face,

The dais at the Los Angeles Press Club, where in 1985 Howard Bennett flushed effigies of LA politicians down the toilet at the Dirty Toilet Awards. *Left to right:* Leif Bennett, Chantal Toporow, Howard Bennett, and Patrick Wall. Photo courtesy of Howard Bennett.

"Thank you very much for sending me the announcement of your news conference regarding Mayor Bradley's role in allowing the Los Angeles City Sewer System to deteriorate. He has earned the 'honors' of the 'toilet award' through years of neglect of the city's infrastructure while currying the favor of the unions with large salary increases. The Board of Supervisors meeting prevents me from attending your news conference. You have my best wishes for continued success in your efforts to keep the ocean clean and viable."

Bennett collected three guests to face the cameras and support his cause, making this appear a bit more official or at least less crackpot. His son, Leif, wearing a sweater and open shirt, sat to his father's far right, looking a little sheepish, as though he wanted to hide behind the drapes that surrounded the room. When a detail-oriented reporter asked who he was, he identified himself as a coalition member, that is, the vice president of the Southern California Jujitsu Association. The implication was: No, I'm not a scientist or anyone who actually has any expertise in this. Without intention, no doubt, the toilet had been placed next to him, as if to confirm his lack of credentials.

Perhaps to counterbalance his bench-warming son, Bennett sat Chantal Toporow, a real live scientist, next in the lineup. Patrick Wall took the fourth chair, representing Earth Alert!, an environmental organization that he and his wife, Janet Bridgers, had founded the year before. Wall had shown up at the brown ribbon demonstration wearing a Neptune costume, but he came to the awards in a sport coat and tie. He was also the one person in the room with the kind of enviro chops few activists could equal—he had gone into eco combat for Greenpeace by challenging hunters who were out to kill baby seals and, in one internationally famous instance in 1980, going to jail after freeing dolphins trapped in nets by Japanese fishermen. Behind the group were two easels with press clippings regarding the coalition, Hyperion, and Santa Monica Bay, which to Bennett's mind added the necessary cachet to his enterprise, suggesting that, if the topic was good enough for the *Los Angeles Times* et al., it was certainly worthy of the folks in the room.

Before the flushing began, however, Bennett spiced the occasion with his usual vitriol. "The Dirty Toilet Award," he growled, "is given to Mayor Bradley and the Los Angeles City Council for using Hyperion Sewage Plant to destroy Santa Monica Bay. They have been dumping into Santa Monica Bay as if it were a dirty toilet for almost thirty years. . . . The Dirty Toilet Award

Howard Bennett conducting the awards presentation before the press, 1985. Photo courtesy of Howard Bennett.

is given to Mayor Bradley and the Council for a continuing pattern of a 'we don't care' approach to the people. When 100,000 gallons of raw sewage spilled from Ballona Creek on September 21st, the public was not given a chance to save itself. The City didn't even call the lifeguards and tell them to get people out of the water or to stop fishing in raw sewage! . . . Here again, it's a switch on the saying 'kiss and tell.' Los Angeles has a policy of 'Dump in the tide and hide.'"

Bennett rolled on, slamming Bradley and the city over and over, demonstrating his flair for bright, shiny sentences glowing with acerbic accusations. The summer vacation had done him good, and after more than six months of campaigning, Bennett had honed his act to the point where the lines flew out of his mouth like arrows. He remembers the reporters laughing, if not a little enthralled with the performance, but reporter Marc Haefele recalls "nervous laughter and embarrassment." As he puts it now, "I thought [the awards] were a little far out at the time. I thought, he's keeping the franchise alive and that's good, too, but this is a little less to the point than his earlier disclosures." Or, as a KCOP-TV reporter put it during a happy-talk segment before her story about the awards, "You might as well put a little humor in your political activism."

Bennett concluded with an overblown promise: "The Dirty Toilet Awards will keep coming. It would be a foolish politician indeed who thinks the voters in their area at the next election won't want to know if he or she got any Dirty Toilet Awards. If the answer is 'yes,' the next question will be 'how many?' The politicians can be sure—we will tell them!"

Still on fire, he walked to the awards hung on an easel, and the presentations began. Each target of Bennett's wrath was represented by a gingerbread-man-like cutout, faceless and plain, made of blue paper. One by one, Leif stiffly held the effigies above the toilet bowl, festooned with the words "STOP DUMPING INTO SANTA MONICA BAY," and dropped them out of sight. It didn't take much imagination to picture what would happen next were this a real toilet connected by miles of pipe to Hyperion.

CHAPTER 14

The Decision

In early 1964, Los Angeles installed an odd contraption on concrete-lined Ballona Creek at Jackson Avenue in Culver City that largely went unseen for some twenty years. Spending more than $161,000, engineers built a hundred-foot-long, six-foot-wide reinforced concrete pipe bypass from the main sewer line to an eight-foot-by-ten-foot concrete box. At the time, this must have seemed like a simple solution to one of the city's oldest problems: sewage spills when volume exceeded capacity in the sewer line, usually during heavy rainfall. Incredibly, there were times in the early twentieth century when raw sewage burst from sewage lines and literally ran down the street in what can only be imagined as one of the most noxious accidents known to urban life.

At the Jackson Avenue overflow box at times of such over-capacity, gates opened and the flow filled the box until the wooden lid, sliding up and down on six rails, popped open and the sewage tumbled out into the creek. A crew was then dis-patched to the site to shovel chlorine on the resulting mess, assuming they got there before it poured down the trough in the middle of the flood control channel that is the "creek."

To those who designed the overflow box, it made more sense to spill a few gallons of sewage—which eventually ended up in the Playa del Rey surf—than risk an entire sewer line bursting from the excess pressure and losing hundreds or thousands of gallons instead, perhaps on some suburban street. Well, maybe. But on July 20, 1985, their solution came back to bite the city in its big ol' concrete butt.

According to Heal the Bay lore, Dorothy Green's brother, Jerry Cohen, happened to be working that day, a Saturday, on an industrial building under construction at the family's property near Ballona Creek. In the kind of almost comic cosmic coincidence that the Howard Bennetts of the world see as pure justice—because of how the sewage spills finally gained public attention—the overflow box's lid popped loose so violently that a glob or two of raw sewage carried by a strong wind landed on Cohen. Putting aside the fact that the overflow usually only dribbled out, and that the lid itself should have prevented wastewater from flying up into the air, this makes a great story worthy of Bennett himself, however apocryphal some of its details might be. Cohen called his sister, the hotshot enviro, and asked her for an explanation.

For his part, Bennett disputes this evocative tale as the genesis for the overflow box's discovery. He now says that Jack Green, Dorothy's husband, found the contraption near the construction site by a means far more mundane—the stench that occasionally wafted over the area after a spill. Jack showed Bennett the box, believing it might be pertinent to his campaign, but Bennett doesn't recall exactly when this happened, guessing it must have been before July. (Jack Green died in 2005, before research on this book began.)

However Dorothy Green was alerted to the overflow box, she eventually learned that the spill that supposedly splattered her brother wasn't all that rare. In an eleven-day period starting July 12, the box gurgled forth a total of 10,000 gallons on four different occasions. This was a mere drip compared to August 17, 1977, when 2.4 million gallons of raw sewage popped loose from the overflow box.

As though the coincidences couldn't get any better, a day before the first spill, Robert Ghirelli, the Regional Water Quality Control Board's executive officer, told the *Easy Reader*, a Los Angeles tabloid newspaper, that the 301(h) waiver was likely sailing through the approval process. The article dryly noted that attorney Felicia Marcus of the Center for Law in the Public Interest and the Sierra Club's Nancy Taylor were locked and loaded for just such a possibility, ready to march the EPA and water board into court. But, Green discovered, once that overflow box burped, the work had been done for them.

According to former assemblyman Tom Hayden, a staffer showed him the Jackson Avenue overflow box just prior to the July spills. "And around the box, a lot of weird things were growing—flowers and everything," he says now. "The kind of stuff that pops up when the sewage sinks into the soil. It looked like a jungle, a small minijungle. And that was it for me. That was the beginning. That was the realization that the drainage of all of LA was not built to handle runoff when it rains."

He announced his discovery in early July and his belief that the spills were going unreported by Hyperion officials. "This is a crime," he said with a certain indignant flourish. "Somebody ought to be held accountable."

Green remembered a press conference at the box itself, where she and Hayden shared the chastising duties. A news videographer, she said, pointed out a tomato plant growing nearby, the indestructible seed from which it grew having traveled through someone's digestive tract and then the city's pipes before landing there.

The publicity over the spills quickly reached the Regional Water Quality Control Board, where, in early August, Ghirelli fined the city $30,050 for violating its wastewater discharge permit. It should be noted that the permit actually allowed spills during the wet season when heavy rainstorms could overload the system. The fines in question, however, regarded so-called dry-season spills, which were prohibited. But given that this was the first time the city had been pinched for its many sewage spills over the years, it's clear the state hadn't been paying much attention until now.

Despite what he might have said the month before regarding the 301(h) waiver, Ghirelli now says the spills changed the board's thinking. "Because there were so many problems occurring [with the infrastructure]. How could you grant a waiver to the city when they couldn't handle their sewage system? That to me was the defining moment, if you will."

The fine certainly influenced Mayor Tom Bradley. A month later, on September 4, he sent a letter to the water board's chairman, James Grossman. While not exactly contrite about the past years, Bradley dropped this shocker: "To achieve our goal of the cleanest possible bay, we are *determined* [the emphasis was his] to install a full secondary treatment system at the Hyperion Waste Treatment Plant."

Then again, maybe Bradley wasn't as committed to that *determination* as the opening paragraph suggested (newspaper articles quoted only those lines). Later in the letter he back-tracked and wrote, "For Los Angeles, there is only one question: How can we install a full secondary treatment system at Hyperion? After a thorough analysis of all available data, I have concluded that the answer to this question is clear: The fastest, most cost-effective way to achieve full secondary treatment at Hyperion is for the City of Los Angeles to begin *immediate construction* [his emphasis] of the partial secondary treatment system."

He touted partial secondary treatment because, as he admitted in the next paragraphs, the city already had the money in place. "All that is required is your Board's approval of our Section 301(h) application," he wrote somewhat disingenuously. In other words, first give us the waiver, and we'll get right on partial secondary treatment, but, uh, full secondary will have to wait.

Unfortunately, what Bradley—or more precisely, his staff—didn't understand was that, in order to qualify for the federal and state funding available for the Hyperion construction, they had to withdraw their waiver application and commit to full secondary treatment as required by law. They couldn't have it both ways.

Finally, Bradley's letter ended on this note: "Once the initial construction is completed, I am determined to see that construction on a full secondary system begins as soon as possible." Forgetting for a moment that this equivocation never made it to the public, there were three guesses at the time about why Bradley had changed his mind, sort of. Hayden surmised it was the summer's protests. "In the face of a very embarrassing charge the city is polluting the bay," he told the *Times*, "he is at last taking steps that should have been taken before, but they are welcome steps."

From Grossman's perspective, the water board's fines influenced the mayor. "I think it was embarrassing," he said in the same *Times* article. Or perhaps Bradley was close to announcing his second campaign run for governor—a 1986 rematch with Governor George Deukmejian, who barely won the first time—and realized his environmental record needed some sprucing up.

Unfortunately for Bradley, three weeks later, on September 21, the overflowing sewers this time disgorged the worst spill yet that season, 95,000 gallons of raw sewage, which got past the brigade usually sent to neutralize the sewage with chlorine and flowed into the bay at full strength. Perhaps more galling for Bradley, it took four days for his staff to tell him about it.

"He did a lot of screaming about the sewage spills around city hall, for sure," Maureen Kindel, then the president of the Board of Public Works and ultimately responsible for the spills, says now.

The public didn't hear about the spill, either; Los Angeles County Department of Health Services officials said it didn't result in a bacterial spike at the local beaches, so not to worry. Nevertheless, Kindel ordered the city from that point on to alert the press every time a spill occurred.

It just wasn't looking good for the mayor. And it only got worse on November 25, when the Regional Water Quality Control Board held its final public meeting on the 301(h) waiver. Playing it by the book, the board gave Bennett and any others one more chance to make their case for denying the waiver. It was a strange notion, actually, considering that the board had already unanimously rejected the waiver. Insofar as democracy goes, it was an empty gesture, but it made everyone in State Building Room 1138 feel better.

Wearing a dark sport coat, tie, and khaki pants, Bennett—already knowing the board's decision—put aside his vitriol and stood at the podium a calm, mannered guy, sounding far more adroit on the topic than he had been when all this had started nine months before. Identifying himself as "a teacher at Culver City High School," he said, smiling, that "many of my students have come here today, and they have been actively involved in helping clean up the ocean. They are on the side of the angels." He paused and looked up from his script as though meeting Chairman Grossman's eyes. "As I feel you are, too," he added gently with a mixture of gratitude and relief; he had forgiven them for their previous mistakes.

(The following January, Grossman belatedly excoriated the EPA for tentatively approving the waiver in the first place. "Where the hell did the EPA come out making this recommendation?" he told the *Times*. "How in the hell can I have faith in an agency who did this?")

Rim Fay, on the other hand, wasn't quite as conciliatory as Bennett, and in fact, after so many years of getting nowhere, he displayed the cynicism he had earned. Looking uncharacteristically professorial in a heavy sweater, white shirt, and tie, he said in his usual crisp, resonant voice, "There is no resolve obvious anywhere in the city except in the office of the mayor to support secondary treatment. Not one city council person has spoken for it. There's no word coming out of the Department of Public Works or the Bureau of Sanitation resolved to go ahead and implement this long overdue improvement in the protection of our environment and *our* Santa Monica Bay."

There wasn't quite the same boisterous crowd at this board meeting as at the one held the previous May. Nevertheless, the seats were full and several people held up hand-lettered signs with "Heal the Bay" on them, which at the time appeared to be declarative sentences rather than the name of a largely unpublicized group. If those people had known about the rift between Bennett and Heal the Bay's president, Dorothy Green, they might have also appeared to be mocking Bennett and his only-on-paper coalition.

In a way, however, this was simply a low-key celebration two months after the 301(h) waiver had been unofficially denied. In a September 30, 1985, letter from EPA regional administrator Judith Ayres to Mayor Bradley, the city was informed, "Due to new information available since the tentative decision was issued on November 30, 1981, and technical deficiencies in the tentative decision document, it is our decision to re-evaluate the City's section 301(h) application for the Hyperion Treatment Plant." They were, the letter said, giving the city a time extension, until July 1, 1988, to meet secondary treatment requirements.

For its part, the Regional Water Quality Control Board issued this formal statement three days before the meeting:

> Whereas the Santa Monica Bay is a heavily used body of water which has been found to be stressed due to a large number of wastewater discharges; and
>
> Whereas the applicant has failed to demonstrate that the discharge from Hyperion Plant will not result in degradation and further stresses of marine communities, including vertebrates and invertebrates; and

Whereas the application has failed to demonstrate that the discharge from Hyperion Treatment Plant of less than secondary treated effluent will not contribute to an increase of waste materials in ocean bottom sediments; and

Whereas the discharge from Hyperion Treatment Plant of less than full secondary treated effluent will not protect the beneficial uses of Santa Monica Bay and will not maintain an indigenous marine life and a healthy and diverse marine community;

Therefore be it resolved that the California Regional Water Quality Control Board, Los Angeles Region, does not concur with the issuance of a waiver from secondary treatment at the Hyperion Wastewater Treatment Plant.

In announcing this at the public meeting, Chairman Grossman sounded a little weary of all that had preceded this day. "I think this board wants to work with the city, and we want to try and get out of this political morass that we keep getting pushed into, and I think we will get out of it. Um, if necessary, however, we will talk about moratoriums on hook-ups. We will talk about cease and desist orders." This might have been a curt nod to Fay, who had insisted for months, if not years, that the city halt new construction and its resulting increase in sewage until it beefed up Hyperion.

According to attorney Felicia Marcus, who by this point had joined Heal the Bay as its counsel, Grossman had other influences tugging on him. "Afterwards in the hallway," she says, "I was talking to Grossman. . . . It's, like, his nine-year-old daughter, who said to him, 'You're not going to let them keep doing this?' And clearly the nine-year-old daughter knew because of all the publicity and all the hullabaloo; otherwise why would she know?"

Following the meeting, Bennett, looking humble and sounding uncharacteristically slow in choosing his words, told a television news reporter, "I can't believe it, but it's a victory for the people. And I think that this is the beginning of a rebirth of Santa Monica Bay. And really, I congratulate everybody who helped, which has been everybody." It all sounded so good, but Rim Fay was right. This didn't end the matter.

CHAPTER 15

Friend of the Court

Felicia Marcus was waylaid by sewage. Sewage became her life, her obsession, the focus of her goals. It was entirely possible that even the Hyperion engineers weren't as enthusiastic about the subject of treating what people flushed down the city's pipes as she was.

Sewage was—and is, of course—the most basic of a society's challenges. A city as large as Los Angeles produced enough sewage in one day—420 million gallons in 1985—to create the state's tenth-largest river (metaphorically, at least). And how to deal with that flood of effluent fascinated Marcus. In 1985, when she was twenty-nine, her career goals suddenly shifted to making sure the city never polluted the Santa Monica Bay again with human waste and chemicals.

Sure, she could understand what made people recoil if she brought up the subject, say, at a party. But cleaning wastewater was so central to the city's survival that they should have been more aware of what happened postflush. At least for a few informative minutes, watching her eyes brighten, hearing her voice get a little louder and even cheerful as she rapidly described, say, the intricacies of secondary treatment, one might temporarily find a soft spot for the subject of sewage treatment.

Or, at the very least, her listeners surely found a soft spot for the effluent evangelist herself, even after saying "Yuck" a few times. It seems few people don't love Felicia Marcus. She is enthusiastic, bright, and attractive. While other enviros at the time could be strident pains—and, to be sure, Marcus had just as much fire for the cause as they did—she found ways to downplay the differences between her and those she opposed. Instead of feeling an adversarial push from her, they got a nearly subversive sympathy for the tough jobs they had. After all, she was dealing not with criminals but with engineers and bureaucrats, and even if guys like Howard Bennett figured these people had botched their jobs, Marcus smiled and, instead of vilifying them, essentially said, "I know you're trying to do the right thing. Let's work on this together." And with few exceptions, they not only responded to the respect, they *liked* her.

At the time the Regional Water Quality Control Board denied the city's 301(h) waiver application, Marcus still hadn't completely refocused her law career on sewage, although she was definitely involved in the Hyperion issue. With Heal the Bay just getting started, she joined the group about the same time that she became involved with the Los Angeles League of Conservation Voters' board, acting as Heal the Bay's counsel, able to interpret EPA documents and other legalese that flummoxed her fellow enviros. This more or less fit in with her ultimate career goal, which was to become an eco prosecutor going after individuals and companies that broke environmental laws. And

Felicia Marcus in 2008. The attorney injected Heal the Bay into the
consent decree's negotiations concerning the Hyperion plant and
later ran the city's Board of Public Works.

we're not talking about just squeezing some money out of them with fines, but tossing the perpetrators in jail. She wanted these guys to hurt a little for their sins.

But then sewage got in the way.

Marcus grew up with her aunt and uncle in a comfortable hillside home overlooking the San Fernando Valley. After graduating from Harvard in 1977 with a degree in East Asian studies, she eventually worked as a legislative assistant for then Representative Anthony Beilenson, a Democrat from Los Angeles' Westside. Even though at first Marcus considered environmentalists a shade elitist, she eventually grabbed the office's vacant eco slot and was hooked by the seminal legislation running through Congress at the time, including the Superfund act in 1980 and the Alaska lands act the same year, which set aside huge land tracts for conservation. Around the same time, she first met, by phone, Dorothy Green, who at that point was crusading for changes in California's water rate system.

After realizing that so much of environmental activism involved the legal system, she enrolled in New York University's law school, paying for it with a Root-Tilden Scholarship and specializing in public interest law. From there, she returned to Los Angeles to clerk for Judge Harry Pregerson of the Ninth Circuit Court of Appeals, a position that later had some significance in the Hyperion saga. After that, she became a visiting fellow at the Center for Law in the Public Interest.

That's when she got her first whiff of L.A.'s sewage issues. After the March 25, 1985, Regional Water Quality Control Board public hearing, where only a few people testified against the 301(h) waiver, Larry Lacombe, an activist from the Sierra Club, asked for a meeting to see what the Center could do regarding the waiver, which at that time seemed certain to win approval. This isn't to say the Sierra Club itself was necessarily interested in the topic. Howard Bennett had called the organization's Angeles Chapter for its support, and they turned him down. Those in the chapter at the time say the club as a whole was more interested in the land and water issues east of the beach and hadn't yet caught on to the bay's pollution.

That same disinterest in sewage was apparent at the Center, so Marcus, at least a little curious, took the meeting, and her fascination began in earnest. Because of Lacombe, she attended the May 13 hearing, met Green in person for the first time, and read Green's testimony to the water board. She also met Bennett and, three weeks later, sent a thorough sixteen-page letter to the EPA and Regional Water Quality Control Board that she cowrote on behalf of Bennett's coalition. Veering from Bennett's sometimes overwrought arguments over public health (the Clean Water Act, after all, didn't concern itself with public health but with the welfare of marine life), the letter methodically attacked the issue from a legal standpoint, not surprisingly quoting the act's statutory requirement that a waiver not interfere with the "balanced indigenous population," and then demonstrating how the bay's BIP had already been compromised. She and coauthor Joel Reynolds concluded by pointing out that "the requested waiver is legally unsupportable."

Once her fellowship ended, Marcus picked up an associate job at Munger, Tolles, and Olson, a law firm with a reputation for allowing its attorneys the freedom to perform pro bono work with the office's resources provided they kept up with paying clients as well. She quickly realized that the truly interesting legal work there was the free stuff, and, while the paycheck was nice, she knew she couldn't stay there a long time. It would be boring. Here, Marcus worked on Heal the Bay activities, providing the legal advice Green needed.

This included accompanying Green and others when they called on Mayor Bradley's office sometime during the autumn of 1985—and before the final waiver denial—for a what-are-you-going-to-do-about-it? session on secondary treatment at Hyperion. Given that Heal the Bay was, strictly speaking, just a few people upset about pollution in the bay, it's remarkable that Bradley's deputy mayor, Tom Houston, even took their call. But he did, no doubt recognizing Green's name from her previous environmental causes. A small group including Marcus, Moe Stavnezer, and Green entered Houston's office anxious for this momentary brush with the city power structure. It didn't last long.

"We went in," Marcus says now, "and very logically laid out this whole case, and he did basically tell us to go to hell. He didn't use those words exactly, I don't think. But basically [he] said no way anyone would vote for the bonds [to pay for the Hyperion upgrade to full secondary treatment], essentially, and just didn't care."

Again, this may have been a case of Bradley's staff insulating him from the issue. As Maureen Kindel, then president of the Board of Public Works, puts it now, "I don't think [Houston] was very helpful. . . . It did not seem to be an issue that turned him on very much."

Still, given both Green's and Marcus's belief that you don't publicly out officials for rude behavior, but keep working on them, they kept quiet. Green even went so far as to say she didn't remember the incident. "That kind of thing rolls off my back," she said. While this refusal to indulge in the public chastising of city officials may have kept Heal the Bay out of the newspapers, two events were coming up that would turn the tiny group into the major player Green and Marcus both wanted it to be.

The first incident was brought about by Kindel, who had her own philosophy of how to get along with people. "Life is still about relationships," she says now. "And I knew what had to happen was to bring people together. And to informally—not formally, informally—discuss this. And I also had to set a very clear direction to the professionals who worked for me at the Department of Public Works, and I had to set that direction with the mayor—with the mayor's approval, I mean. And with the people who worked for him that were advising him on environmental matters. And I had to push my point of view."

So as the sewage spills became hot news, "one day I was just really kind of, sort of, pressed to the end of my endurance," she says. "And I asked for a meeting to take place in the conference room of Public Works. It was a brown-bag lunch, and I wanted everybody who worked for me in a leadership role that was involved in the Hyperion plant—which was enough, believe me—to get the engineers and the Bureau of Sanitation people to sit down together. Already, we're in a very egalitarian mode. A simple thing like that," she adds facetiously. "Can you imagine in my own department getting people to talk to one another?"

Just one problem: the meeting had a distinct aura of secrecy about it. Those attending didn't say much about it outside of their departments. Then again, according to Kindel, it *was* sort of publicly announced, but only on a bulletin board somewhere in the bowels of city hall. However, a member of her staff says that this wasn't really a public meeting, and no notice was given for what was essentially a staff meeting. Just the same, someone there dropped a dime on Green and told her Kindel was set to discuss Hyperion. Given that Green had already sued the city before for holding public meetings without proper notice, she decided to crash the party.

But first she stopped by the press room and let it be known that Kindel was holding a "secret meeting," and that they should find this interesting. The bait attracted several reporters, and they walked to the conference room where Kindel was holding the conference. What happened next both women acknowledge as a crucial moment. To defuse any notion the public had not been invited, Kindel welcomed Green into the room, offered her coffee, and got her a chair against the wall where she could see and hear the proceedings. "But, of course, I was the only one being this way," Kindel says. "Everybody else treated them like somebody's dirty shirt had arrived."

For the record, the reporters lost interest fairly fast, but Green stuck around, attending more meetings during the coming weeks and making what she felt was the most important progress she could make—she now had a seat at the table. At first shy, according to Kindel, Green piped up enough to influence the group. Heal the Bay, however small, now had a voice.

The meetings themselves did produce one important decision. Kindel hired Don Smith of James M. Montgomery Consulting Engineers, a Pasadena-based engineering consulting firm that was managing the construction of the Hyperion Energy Recovery System (the process to burn sludge), to oversee the plant's operations (two top plant officials were reassigned) and to also head a group of other consultants who combed Hyperion looking for ways to fix the aging facility. Smith's February 1986 report detailed a sewage treatment plant springing leaks in every possible category. "On an overall plant basis," the two-inch-thick report said as precisely as it could, "significant concerns were noted about the general facilities deterioration, sampling methodology, operation performance monitoring, and odor control. Sampling techniques at the plant need to be thoroughly reviewed and modified to assure that data used to monitor and control process operations are representative of actual plant performance."

Besides the infrastructure problems—including in just one year 493 permit violations related to the plant's five-mile outfall—Smith also found a staff barely working each day, their morale tattered by disinterested bean counters who refused to give them enough money to maintain the plant: "Significant non-structural issues contributing to overall morale problems identified were lack of focus of effort, inadequate inter-sectional cooperation, inadequate communication and limited career path development opportunities."

About two months before this February 1986 report, the EPA had added to Mayor Tom Bradley's apparent image as the guy who had polluted the city's perfectly good beaches by announcing it would study Santa Monica Bay as a potential Superfund site. Given that this designation usually went to toxic dump sites—and what could be worse than the 6.6-square-mile DDT patch on the bay's floor?—it made the bay appear so contaminated that even swimming in it held a carcinogenic risk. This didn't help Bradley's shot at being elected governor in November 1986. (To be fair, the DDT had come from the Los Angeles County sewage treatment plant, not Hyperion, and therefore Bradley shouldn't have been held responsible for that particular pollution. But in politics, fairness isn't always considered.)

The next day, December 17, 1985, the Los Angeles City Council hid away in closed session (so they could discuss legal issues out of the public's earshot) and emerged with this decision: They would comply with the water board's order for full secondary treatment at Hyperion. Keep in mind, this was mostly the same bunch who, years before, voiced by Zev Yaroslavsky, had advocated dumping sludge into Santa Monica Bay. As city council president Pat Russell put it to a *Times* reporter following the closed session—no doubt through clenched teeth—"It's the right, honorable thing to do."

The right thing was going to cost $528 million as part of an overall $1.9 billion wastewater improvement program (the figure jumped to $2.6 billion a few months later). Unfortunately, the federal funds promised under the Clean Water Act to help cities meet the law's sewage treatment standards had gone dry after the Reagan administration's EPA director, Anne Gorsuch, cut the agency's budget by 22 percent. So the city predicted that residential sewer rates would eventually triple, from the current $5.40 a month, to help pay the bill.

It seemed apparent once the 301(h) waiver had been denied that most of the heavy lifting was accomplished. But the EPA still had to take care of some legal business. The agency reopened a years-old lawsuit it had filed against Los Angeles to force the city to abide by the Clean Water Act's requirement for full secondary treatment. The EPA had put the lawsuit on hold once Los Angeles filed its 301(h) waiver application. Now that the waiver issue had been settled, the lawsuit was resurrected, and it landed in the courtroom of Ninth Circuit Court of Appeals judge Harry Pregerson, for whom Marcus had worked a couple of years before.

The sixty-three-year-old Pregerson had a reputation for leaning toward the green side of things, and the enviros who sat in the courtroom felt that the judge would inevitably support their point of view. "I love that man," Marcus says now. "Talk about a people person. That guy gets totally how the world works, as opposed to being stuck in the pinheaded—he knows the law . . . but he'll do everything to cajole, wheedle, and try to get somebody to do what he wants to do. He's brilliant."

In a hearing in late May 1986, several groups clamored for a slice of the litigation pie, all of them afraid that the city would bang out an agreement with the feds that didn't immediately force Hyperion to stop discharging sludge into the bay—it had just missed another sludge-out deadline the previous February—and which didn't hold all parties accountable. Several seemingly disparate organizations banded together—the Fund for the Environment, the Marina del Rey Anglers, Harry's Bait and Tackle, and the Los Angeles County Lifeguard Association—and won the right to intervene later, that is, to file an appeal, if the EPA and state were, in their opinion, too lenient on Los Angeles.

Marcus looked at this and figured her client, Heal the Bay, deserved the same status, so she simply raised her hand to be heard. "I didn't think we were going to need to appeal his decision," she says now. "He was very green. [But] I said, 'Can you make that amicus [curiae] with leave to intervene and add Heal the Bay and California Environmental Trust? [another group Marcus represented],' and he said, 'Sure.' I think he even had me draft the order, because he can just ask someone to draft the order. I can't remember. It was just very informal in a way . . . but that's not unusual for him."

With amicus curiae, or "friend of the court," status, Heal the Bay, a group with few members and, up until this point, little influence, quickly inserted itself into the legal proceedings. Marcus made the most of her position. During the first meetings, where the various litigants and friends of the court toured the Hyperion Sewage Treatment Plant, city engineer Bob Horii came under scrutiny for the ways things had been done up to that point. The frustrated lawyers in the room believed he was dodging their questions, not realizing that engineers can sometimes be a little too precise for their own good.

"So partially just to amuse myself, maybe," Marcus says now, "and maybe even [because] I'm not the nicest person in the world—I've become way nicer because of this experience and more empathetic—I started asking the same question. [She doesn't remember now what the question was.] I may have asked it ten times, and each time I asked it a slightly different way, like an experiment. . . . I finally asked it a certain way, and it was, like, boom, the cash register opened. And he answered what everybody had been asking. . . . He hadn't answered [earlier] because we hadn't asked it a certain way. . . . It was: Oh, my God, it's not that they're trying to keep it from us, it's that they speak a different language."

As the interviews went on, Marcus discovered that the men so many people had vilified were just a bunch of guys trying to do their jobs. She turned to Harry Sizemore, the plant's manager. He admitted he wouldn't let his daughter swim in the bay (he had publicly said the opposite—recall that he defended the plant in 1985, saying it was doing a good job). As Marcus says, "He was an honest guy, and he explained to us how it [previously] had gotten that way. . . . They had a sanitation director who really didn't believe in fixing it up, for many years." She developed a certain affection for the guys running Hyperion and sympathized with them as no other activist had done. While they may have spouted the company line in public, Marcus, with her disarming manner, found them to be honest and willing to reveal their troubles in private.

About a month after this lesson in humanity, the parties finalized a consent decree, and the Los Angeles City Council initialed it on July 30, although it didn't become official until Pregerson approved the settlement in February 1987. With that, the city agreed to stop dumping sludge by the end of 1988 and then, by 1998, to finish installing full secondary treatment. Marcus and her fellow activists had asked for an earlier deadline; but after the city had argued that the construction couldn't go as fast as the environmental groups would have liked, Pregerson said no. The enviros had also called for an independent monitor to watch over the city's progress, and Pregerson decided instead that he could monitor the work himself. However, he agreed that the groups could request progress reports. "So essentially what I negotiated for was Heal the Bay being in their face for twelve years," Marcus says. This meant that Marcus and her cohorts could review the city's efforts with the judge once a quarter; she believed this kind of scrutiny would give Los Angeles an incentive to meet the consent decree's deadline.

Heal the Bay, with Green claiming a membership of nine hundred individuals and sixty organizations, was now more than a bunch of protesters; it was a part of the process. As

Green told the *Los Angeles Times*, in the kind of statement that might be considered elitist elsewhere but was a declaration of credibility in Los Angeles, "We have lawyers, film industry people, media industry people: they are just coming out of the woodwork."

On the other hand, Howard Bennett's coalition had lost much of its profile as Heal the Bay took over the pollution issue. However, with the consent decree, Bennett found one last opportunity to rage against the city to a reporter from a tiny newspaper, the *Culver City Wave*. "[The consent decree] is a dirty deal . . . ," he said, "dirtier than the toxic sewage dumped into the bay." And with his customary cynicism, he added, "Deadlines to the city are meaningless. They have connived and cheated to get out of them."

Marcus, however, wasn't unhappy with the compromise. "This is part of the whole philosophy that Dorothy let me really run with on the issue side, which was: Let's be as tough as we need to be, but let's also be willing to capture results," she says. "You know: they offer you 80 percent; you take the 80 percent and then you sit down to talk about the other 20. Versus to continue to beat them over the head unless they give you 100 percent, which is a major flaw in the environmental strategy in a lot of places."

In a way, Marcus figured, she had been clever in negotiating to give Heal the Bay the ability to monitor the city's progress. But ironically, three years later, her former cohorts would be watching over *her* shoulder.

While all this took place, Mayor Bradley busied himself as the state's Democratic candidate for governor, running against the incumbent, George Deukmejian. Bradley must have thought he had a good chance of advancing his career, given that the first time the two ran against each other, in 1982, Deukmejian barely squeaked by, with 52,295 more votes out of 7.6 million cast. It was so close that the once-favored Bradley at first refused to concede.

It might have been even closer had Bradley been more politically cognizant of the state's environmental groups, especially in the north. They turned up their noses at both candidates after the two men supported building the Peripheral Canal, a water supply project that would have ensured Southern California received a portion of Northern California water. The state's largest environmental organization, the Sierra Club, refused to endorse Bradley in 1982. (Dorothy Green worked with the opposition to the canal.)

In 1986, Bradley realized he needed those green votes and, perhaps for the first time, started showing a more public interest in the environment. He conceded that Southern California shouldn't get so greedy over the water supply and pledged that, before the area received any water from neighboring northern counties, it would first store more water and increase conservation. And two months before the election, he took a swing at one of the great symbols of Los Angeles' water imperialism, Mono Lake, a highly saline jewel just east of Yosemite National Park in the northern Owens Valley. Los Angeles had been piping water from the Owens Valley since the early part of the twentieth century and diverting so much *agua* that the lake was drying up. Bradley called for LA to allow the flow to Mono Lake to increase, which the city's own Department of Water and Power opposed. Soon after he made this pronouncement, the Sierra Club rewarded him with their endorsement.

Just the same, Bradley's vulnerabilities remained Santa Monica Bay and the impression many people had—especially in Southern California—that he had resisted ending the city's practice of dumping sludge and partially treated sewage into the water.

With that in mind, someone from the Deukmejian campaign contacted Howard Bennett. As he tells the story: "He says, 'Mr. Bennett, we are the Republican Party, yadda, yadda, yadda. A lot of comments that you have made about the City of Los Angeles and the mayor—Mayor Bradley—are you a Republican or a Democrat?' I said, 'I'm neither, I'm for the ocean.' As a matter of fact, my family had voted Democratic through the years. But what the hell. So he says, 'Would you mind if we use your statements during our campaign for our candidate against Mayor Bradley?' I said, 'Be my guest.'"

The resulting radio ads excoriated Bradley for his environmental record. In one, with the sound of gurgling sewage in the background, a voice-over explains:

> In the next 60 seconds, the largest polluter in California will dump nearly 300,000 gallons of waste into Santa Monica Bay, containing a half dozen toxic compounds known or suspected to cause cancer. And this illegal dumping goes on all day and all night—148 billion gallons a year—in addition to fifty thousand tons of sewer sludge.
>
> Can you guess who's responsible for California's largest pollution problem? Oil companies? No, it's Tom Bradley. That's right . . . Tom Bradley. For the past nine years, Bradley's failed leadership has resulted in the violation of

federal law and the repeated dumping of raw sewage into waters where people swim and fish.

Because of Tom Bradley's mismanagement, Los Angeles taxpayers have had to pay hundreds of thousands of dollars in fines, and the sewer rates will double.

Think about it. Tom Bradley's running for governor. But if he cares so little about the purity of water in his own backyard, how much would he care about California's environment?

The ad was basically true; however, it did overstate the notion that all this sewage discharge was "illegal dumping." Even though the sludge going out the seven-mile pipe violated the Clean Water Act, the EPA had given the city numerous time extensions to stop the flow, so technically the city had been granted a temporary exception to the law. And the rest of the so-called dumping came under the city's NPDES permit, thus making it legal. The ad also implied the discharge was raw sewage, not mentioning that, in fact, only the spills had been raw sewage, and that the fines were for those transgressions, not anything the Hyperion plant itself was doing.

The Deukmejian campaign cranked out eight versions of the same basic message, but the others related to other environmental sore spots, such as San Francisco Bay and drinking water. It should be noted that none of them mentioned Bennett.

While Bradley's defeat that November can't be blamed entirely on Santa Monica Bay, Deukmejian no doubt picked up a great many supporters as a result of the ads. He slammed Bradley by winning more than 60 percent of the total votes to Bradley's 37.

Bennett, who still likes to think he had a hand in this defeat, sent out a press release at the time (on the coalition letterhead listing twenty-six member organizations, including the Los Angeles League of Conservation Voters) declaring, "Bradley ignored the environment of Santa Monica Bay, and its toxics and sewage issues crushed him."

CHAPTER 16

Outsiders and Insiders

There are times when you have to ask the inevitable question that comes with an inevitable answer. You have no choice. Your future has already been decided, and you can't do a thing about it. Dave Brown was in that position on February 19, 1987. Sitting across from Jack Anderson, his SCCWRP boss for more than a year, he asked him, "Would it be in my best interest to look at pursuing my career elsewhere?" Anderson told him yes.

Granted, some of us might have put the question a little differently, with perhaps a small, sincere quiver in the voice, but Brown was a scientist, after all, and such formal sentence constructions came naturally to him. Plus, he had a feeling his termination was coming anyway. An easy conclusion, given that in the SCCWRP sphere he was recognized as the guy who blew Los Angeles' chances for the 301(h) waiver and sent the respected Willard Bascom to an ignominious early retirement.

If it looked as though Anderson's assessment of Brown's future at SCCWRP was retribution for all that had occurred,

Brown thought so, too. Ever since he had returned from his monthlong "vacation" mandated after the blue-ribbon panel had exonerated Bascom in May 1985, Brown had been the invisible man, a pariah to most of his fellow staffers.

"I was really very alone," Brown says now, "and much of an outcast." He could have left; he did get job offers "that I walked away from, which probably I shouldn't have, in retrospect. But I was a nerd scientist. I really wanted to learn molecular biology." And, despite the ostracism, SCCWRP offered him that opportunity, at least for a little while.

The conversation with Anderson arose out of Brown's failure to get a grant from the National Institutes of Health to continue his biochemistry work at SCCWRP. (At the time, Brown's research was being paid for by a three-year, $300,000 EPA grant, which ran out on March 21.) When told of the rejection, Anderson let Brown know he didn't want to spend any more of the sponsors' money on biochemistry programs (that is, funds from the City of Los Angeles and the County of Los Angeles, among other government entities that had contracts with the research group).

The next day, Anderson made Brown's termination official after crying poverty. SCCWRP was $109,000 in the hole, he said, and eliminating Brown's position would go a long way toward dealing with the shortfall.

This confirmed Brown's suspicion that he was getting the bum's rush. After all, the administrative officer had told him a month before that he could go ahead with buying a house because SCCWRP's finances were sound. Plus, Anderson had hired someone the previous summer and handed fat raises to most of the staff. He hardly acted as though the books were in the red. Indeed, the same administrative officer later told Brown she had managed to squeeze $70,000 from entities owing them money, and with this and other methods she could keep them fiscally healthy enough to prevent any layoffs.

Obviously, there was more going on, and Brown believed the County of Los Angeles had pressed Anderson to toss him from the program before he could jeopardize another 301(h) waiver, namely theirs, for the Carson Sewage Treatment Plant. As Brown put it in his journal on February 24, 1987, "Jack told several of us during a lunch hour conversation that [Charles] Carry [the Los Angeles County Sanitation District's chief engineer] had called him and said he heard a rumor that he (Carry) was issuing 'idle threats' that we wouldn't be refunded if they didn't get a waiver. Carry then told Jack that he didn't regard them as idle threats, he meant it, we would lose our funding if they didn't get a waiver."

"Certainly the people at our agency were furious with [Brown]," Robert Miele, who worked under Carry, says now. "But I would talk to him, and I thought he was a thoughtful scientist who wanted to do the work he was doing and didn't want it all dragged into the public arena. There were others who said, 'No, he's loving this. He's showboating.' But I never had that sense of David Brown."

Brown noted in his journal that, three months before, "a more junior Los Angeles County representative told me that it was the policy of their management, with their engineering mentality, that we wouldn't be refunded when we come up for renewal in June 1988 if they didn't get a waiver. He asked me if it wouldn't be better for me if I would just be quiet and be allowed to continue to do interesting research. I gave my usual reply that the public has a right to know the facts about the contamination of the marine environment."

One of Brown's coworkers, Bruce Thompson, remembers hearing of the threats, and at the same time the meticulous Brown confirmed with other SCCWRP staffers that they knew Carry held the waiver over their financial future. As one person put it, "Everyone knows about the threats"; Brown recorded this in his journal.

Steven Bay, a lab technician at the time, disputes this, saying now he doesn't recall hearing of threats to lay low on information that might damage the county's chances at the waiver: "I can't think of any instance where [someone said,] 'Boy, let's not report that information. That doesn't look good.'"

Despite the secondhand threats Brown had heard, in a March meeting with SCCWRP Carry denied that their funding was tied to the waiver, but implied Brown had been too negative regarding the county sewage treatment plant's impact on marine life in Santa Monica Bay. Brown and other scientists at the meeting contradicted this Bascom-esque notion and bombarded Carry with the facts about what his Carson plant was actually doing to the bay's ecology.

Anderson's claim and Carry's denial did not reduce the impression that Brown was being booted from SCCWRP, so Brown hired an attorney, Barry Groveman, to make it known he believed his termination was connected to his speaking out. About two weeks later, Anderson reversed himself and told Brown he could stay on, and that the biochemistry program would actually flourish. And then the next day, after Anderson said Groveman had called Carry (he had not), he told Brown and his assistant they were both getting the ax, and that he considered the attorney's action "blackmail."

The drama continued like this for months, going back and forth between job security and dismissal. Brown started to feel pressure from his wife, Anne, to lay low before he lost his job for good. "It kind of troubled me," he says. "Everybody wanted me to back off."

Finally on July 1, 1987, Brown did just that, in a way, and handed in his resignation. "There was a whole lot of people not that happy with what I did," he says now. "And I think that's the same way for every whistleblower, in that people start questioning why you did it. . . . You can always pin a whistleblower as a publicity seeker. That's the easiest thing in the world to do, and that kind of detracts from the message, which was, how do you get the story out? And that was the main purpose, to get the story out. And because I was accused of being a publicity seeker, I got defensive and withdrew to show I wasn't a publicity seeker, and just backed way off, and just completely shut down. I had nothing to say about anything. . . . [But] once I left the scene, there was no one [from SCCWRP] to present the facts."

The next year, he went to work in the immunology department of the City of Hope, a hospital in Duarte, California, specializing in cancer care and research. In other words, he shifted from marine-life molecular biology to that of humans.

Rim Fay and his friend Don May were like restless evangelists, not content with simply saving one city's environmental soul. After the EPA denied the City of Los Angeles' 301(h) waiver, the two took their I-remember-when speeches to waiver hearings for other coastal sewage facilities, scolding and cajoling decision makers with their righteous arguments. Indeed, even before Judge Harry Pregerson signed the 1987 consent decree mandating that Hyperion go to full secondary treatment, an antiwaiver momentum crept over the California coast. By the end of 1986, eleven waiver applications had been denied and seventeen municipalities simply withdrew their applications. The EPA granted only two waivers and tentatively approved another. (In 2009, two California dischargers still hold 301[h] waivers—Goleta and Morro Bay.)

Still, one of the most egregious cases—in Fay's and May's eyes—was a pending waiver application for the Los Angeles County Sanitation District's Carson Sewage Treatment Plant, whose managers wanted to continue with primary treatment only. County San—as they were known—even argued with a straight face that they needed the partially treated sewage to cover the DDT layer still on the Santa Monica Bay's floor.

According to Robert Miele, the idea came in 1979 when they were drafting their original waiver application and it was discovered that the suspended solids going into the bay—the leftovers, if you will, from primary treatment that settled out on the seafloor—were effectively covering the DDT. If they went to full secondary treatment, cutting the suspended solids from about eighty milligrams per liter to fifteen milligrams, natural erosion would eventually peel off that layer of muck. Miele says, "Our scientists came to us and said, 'Look, we did this modeling, and it appears that if we have to go to secondary treatment, the DDT is going to start reappearing at the bottom of ocean. It'll work its way back up, and shouldn't we make that another selling point for why we deserve a waiver?'" At the time, the idea was rejected, but during the public scuffle over County San's waiver in the late 1980s, the managers, desperate for anything to sway the EPA in their direction, resurrected the notion that the suspended solids were actually benefiting the environment. "And of course we got buried by the environmentalists," Miele adds.

Without someone such as Brown to set the record straight, Fay and May met such creative end-runs with the kind of theatrics they perhaps had learned from Howard Bennett. At one hearing they brought in a stinking bucket of sludge Fay had scooped off the seabed.

"Oh, man, it really—it smelled like what it was, right?" May says now. "And there was this reporter, and the media was taking pictures of it. And they're trying to explain how this is— sludge is good for the fish, it's good for the ocean. This is actually a benefit, and [Fay] said, 'There it is, guys.'"

Mark Gold, future Heal the Bay executive director, had watched from his seat, a little awed by the show. "I was just sitting here going, 'Oh, my God, this is just against everything we've ever been taught,'" he says now. "We're supposed to be critical, analytical. Theater—check that at the door. To witness that at my first water board meeting was pretty shocking. You know, I was twenty-two or twenty-three at the time, and so for me, seeing what everyone else's reaction was—I remember what my own reaction was. And just feeling sort of uncomfortable by the theater of it, but then Rim, when he spoke about what he had seen—for me that was the most compelling thing. The theater was forgotten at that point."

The stunt worked, sort of. The water board ordered County San to stop discharging sludge, but the county came closer to winning its waiver. "A lot of people have made him a hero whose advice was not followed," Tom Hayden says now.

It took Heal the Bay threatening a lawsuit, in 1989 (backed by the Natural Resources Defense Council's legal department), to force the EPA to deny the waiver. In early 1987, in the middle of the Fay-May road show, Fay ran for a seat on the Los Angeles City Council as a one-issue candidate (the environment, of course). Once again painting himself as a witness to the environment's destruction, he made simple flyers that proclaimed, "He remembers lessons from the past. He understands the present. He will work for the future of this district." The voters—who might have also seen his campaign bio, which, strangely, listed his thirty-eight years as a lifeguard first—went with another, younger newcomer, Ruth Galanter, in the primary. Endorsed by environmentalists, she later won the seat from incumbent Pat Russell.

"No matter how stupid the politicians, every politician can count," Galanter says of the city council. "And I was elected 58 to 42 percent. All they knew about me was, she's for the environment, so instantly they were all for the environment. Now none of them except [councilperson Marvin Braude] knew a thing about it. But [because of my election] they knew they were for it. And that created a lot of room for action."

According to friends, Fay started drinking more. He lost his driver's license after a DUI conviction. His marine specimen business, Pacific Bio-Marine, started suffering, and soon he moved the lab from Venice to an old welding shop in the Inglewood area, near Los Angeles International Airport, and then, several years later, to the Oxnard area, north of Los Angeles. He continued to testify, sometimes riding his bike to hearings if he couldn't catch a ride with someone. But it had already been demonstrated that the battles couldn't be won by a few individuals—no matter how qualified—grumbling their opposition to one environmental defilement or another. Instead, as Howard Bennett had briefly shown, organized groups flooding government meeting rooms were far more influential.

One such group was the Sierra Club's Clean Coastal Waters Task Force, led by a fiery, diminutive woman named Nancy Taylor, who confronted just about every coastal sanitation district that applied for a 301(h) waiver. Having gained her strategies and fighting know-how from running similar campaigns in Florida and from consulting with her dear friend Fay, she picked apart every engineer who dared utter the phrase "Dilution is the solution to pollution."

However, for sheer influence no one could equal Dorothy Green's Heal the Bay, which had found its niche in both educating the public and promoting itself. The group's name and pervasive fishbone logo infiltrated the city and, more to the point, city hall. While its actual membership started in the dozens and later rose to hundreds, the group looked larger, more prominent, more important than its membership roster suggested. And it all happened so quickly.

In 1986, with Fay as a guest speaker, Heal the Bay held a fund-raiser at the home of Cindy Horn, a member of the original core group. This was perhaps the group's first date with celebrities, a relationship that continues to this day, helping to increase not only their cash flow but their visibility. The matchmaker in this case was Horn's husband, Alan, who was chairman and CEO of Embassy Communications, a company co-owned by television producer Norman Lear. Horn invited his friends and business associates, who also happened to be rich, famous, and well connected. According to Jamie Simons, in an example showing how new the Heal the Bay folks were to the idea of putting out their hands for donations, Horn got up to thank the group but couldn't bring himself to solicit any funds. Finally, Lear yelled out, "Alan! Ask for money!" With that, they raised five thousand dollars, enough cash to move their headquarters from Green's spare bedroom to a small, fume-filled office above a Santa Monica rug-cleaning business on a dead-end street. Later that year, they coaxed a vacant store space out of the owners of Santa Monica Place, an indoor mall, and built an ocean-themed museum there that was open during the summer. Volunteers, including Dave and Anne Brown, scrounged materials for displays, painted a mural on the wall, and then staffed the space. In exchange for free use of the space, Heal the Bay promised the property's owners it would hold a party promoting the mall.

More conventionally, the group held press conferences at Santa Monica Pier, often using Tom Hayden as a draw. They sold T-shirts featuring their new fishbone logo, fashioned after a white croaker stripped of its flesh. The fish in the design looked pessimistic about a healing, but the white outline representing its former body implied there was a little hope. However people interpreted the logo, the T-shirts moved so well that the Heal the Bay moniker went citywide, if not international, when tourists snatched them for souvenirs. Heal the Bay's image became so pervasive that it appeared as though the group numbered in the hundreds of thousands.

They held a "Children's March" in 1989, educating four thousand kids as to why the bay and beach needed to stay clean.

A year later, they raffled off 112 donated surfboards decorated by famous and not-so-famous artists.

Recognizing that she needed the credibility of an in-house marine biologist, Green hired a UCLA graduate student, Mark Gold, in 1988 after he started volunteering for the group. Still working on his PhD, Gold became staff scientist, earning twenty-five thousand dollars a year. And, in keeping with Green's general vision to educate the public, in 1990 Gold instituted one of Heal the Bay's most visible tools, an annual beach report card, which was nothing more than culled statistics from government agency reports detailing bacteria levels. But it worked because it was a scientific, data-free way to not only tell the public about the pollution status at sixty Los Angeles County beaches but also get the organization's name out with little effort.

Quickly, local papers picked up on the report cards, no doubt for their simplicity and quotability factor, and published the grades. The report card has since become a weekly occurrence, and thus more timely, covering 517 beaches statewide, but it doesn't receive as much publicity as it once did.

With the report cards, Heal the Bay started to look like the region's major environmental group. Its influence extended to involvement with a state and federal program begun in 1988 called the Santa Monica Bay Restoration Project (the word *project* was later dropped, and it's now known as a commission). Seven years later, the commission completed a plan—with input from Heal the Bay, among other groups and agencies—to "improve water quality, conserve and rehabilitate natural resources, and protect the Bay's benefits and values." According to the commission's update for 2008, of the ninety major categories the plan outlined for action, about half were either completed or on their way to being accomplished.

Besides bringing more and more prestige to the group, Gold also discovered a talent for raising money, learning that, in Los Angeles, all you have to do is get a few celebrities involved and credibility and money will follow. Soon, Heal the Bay had spokespersons such as Martin Short, Ted Danson, and Julia Louis-Dreyfus to draw crowds at benefit dinners and other functions. The checks soon followed.

By 1990, Heal the Bay's membership totaled seventy-five hundred people, who contributed enough money to fuel a $430,000 budget. Recognizing the political power this implied, the year's gubernatorial candidates, Senator Pete Wilson and State Attorney General John Van de Kamp, sought votes at Heal the Bay's annual meeting by outlining their environmental plans for the state.

None of this impressed Fay. "Rim wanted people to donate their time and make it a priority," a friend, Janet Bridgers, says. "Dorothy created a vehicle where they could donate five dollars a year and have a good effect. That's her great gift, of creating that vehicle where so many people could express their concern so affordably. . . . By contrast, Rim wanted you to go to every hearing. Well, that's not convenient for people who work."

While Fay remained on the outside, fading further away, Heal the Bay's Felicia Marcus scored in 1991 what was perhaps the group's greatest infiltration of city politics and government machinery: she got the chance to run the sewer system that fascinated her so much.

"I think she and Dorothy actually had come in to query our sewer system guys on what progress they were making and so on," Mike Gage, Mayor Tom Bradley's deputy mayor at the time, says now. "And I watched the two of them walk through that process. I actually went over to attend it, along with Mark Fabiani, who was then the mayor's counsel. And I watched them walk through it, and they were never antagonistic with the guys. They worked with them. They questioned them. They were very—they helped to illuminate the issues. I could see lights go on occasionally when they went through the process. I remember Mark and I walking back after that meeting, and my commenting to Mark, 'We need to get [Marcus] into city hall.'"

With Bradley's consent, in late 1987 Gage offered Marcus an open seat on the Board of Public Works with a direct path to the board's presidency. She hesitated. True, here was a chance to have the kind of inside influence no other enviro had ever gotten. The most damning, if not accurate, accusation people such as Howard Bennett had made was that the city couldn't be trusted to keep its word on rebuilding Hyperion. With the kind of optimism that comes from barely understanding the nature of inert bureaucracies, Marcus could have the power to make sure Hyperion met its 1998 deadline for full secondary treatment—the same target date she had helped forge. Then again, she still hoped to become an eco cop, prosecuting those who fouled the earth. Despite how deeply she enjoyed sewage as a topic, she told Gage she planned to go on a long vacation and couldn't take the job.

"That may have been the polite excuse," he says.

"I thought it must be a trick," Marcus says now, "and I wasn't mature enough to really think of the power."

But then a year and a half later, in 1989, Gage called again after Bradley asked him to make another offer. Marcus consulted Green. "You'd be crazy if you don't take it," Green said, recognizing just how useful it would be for a Heal the Bay alum to actually run the sewage show.

The sixty-five-thousand-dollar salary she earned at the start equaled her pay at the law firm, but this was far more interesting work. This was sewage! And this was power! She had always been on the outside, fighting to persuade the decision makers to take her side, and now her opinion meant a lot more. Her very signature resulted in action. She had clout. She pronounced the verdicts. She was the undisputed Queen of Sanitation. Los Angeles' pipes were under her cheerful gaze.

Well, sort of. For one thing, running a bureaucracy hardly makes one cheerful, even someone as perky and energetic as Marcus. Besides that, the same engineers who no doubt flipped off Marcus's predecessor, Maureen Kindel, behind her back were largely running Hyperion. The "dilution is the solution" culture still remained. Marcus had one advantage—in her four years of monitoring the consent decree, she had developed good relationships with the engineers who ran the treatment plant. And so when she told them there was now a new philosophy, that she wanted clean water pouring from Hyperion's pipes and nothing less, they took that as a rallying cry to show what they could accomplish. "If you go for the minimum," she says now, "you get nothing other than to stay out of jail. You don't regain faith with the community. You don't get political brownie points. The way to become an environmental hero—more importantly, the way to regain faith with the public—is to get [the water] as clean as you can."

This isn't to say she didn't get support from the one person the engineers might have shown some respect: Mayor Bradley. "He totally backed me on all of that, against people he had known for years," she says now. "Total integrity on my stuff. I was blown away. I didn't expect it. It was really good. And he would laugh. He'd call me, and someone was complaining about me, and he'd say, 'I just need to know. Somebody called and said blah, blah,' and I'd say, 'They're full of shit' [and then explain why]. He would just start laughing, and he'd say, 'Thanks, I just really needed to know.' People said I amused him because I wasn't the typical political appointee."

Then again, the engineers might have been the easy part. Long before Marcus took over as president of the Board of Public Works, the city had committed to building the Hyperion Energy Recovery System, or HERS, as a way of burning the sludge to run electric generators. At the time, it sounded pretty slick—take the very thing making life so difficult for the city and use it to power the sewage treatment plant. Hey, the Japanese were doing it. Why not Los Angeles? (Incidentally, among the earliest and loudest HERS backers was Don May.)

Everyone wanted this to work. It would have made life so much easier if they could simply turn sludge into energy gold. And indeed, HERS produced 450 million kilowatt-hours of electrical power between 1987 and 1992. Unfortunately, HERS was buggy. "The concentrating system didn't work," Marcus says now of the device designed to wring water from the sludge.

"Part of it is [that] it got gunked up with—there's more oil in our diet than the Japanese. It's disgusting, but it's true." Also, hair in the sludge clogged the system. Sand leaked in through old pipes under the beach, infiltrated the sludge, and burned out the machinery. Marcus says that, despite all the problems, the system did work, but it was so problematic they eventually shut it down, in 1992. "It was technically called a failure so that the EPA would forgive the grants they gave," she says. "But it did work." Then again, it may have worked, Marcus says, but the system took so much effort to maintain that it was no longer practical, and her staff found cheaper and more effective alternatives for drying the sludge.

There were other troubles. The city instituted a water conservation program, partly to reduce the volume going to Hyperion during the construction period. It worked so well that the sulfur in the sewage became more concentrated, which eroded the tiles lining the insides of the sixty-year-old pipes, causing them to drop off and clog the lines.

Still, Marcus pushed the engineers toward a goal no one expected them to meet—instituting full secondary treatment *before* the actual secondary treatment apparatus had been completed by the 1998 deadline. She says this was, in fact, Gage's idea—to make the water pouring out of Hyperion as clean as possible, as fast as possible. The only way to do that was with an expensive system called advanced primary, which used chemicals to clean wastewater beyond what the primary treatment did and equaled what secondary treatment could do. "We broke all the records for what you could do with advanced primary," Marcus says. "It was unbelievable. . . . What the City of LA did—and I only take credit for being the catalyst and then the cheerleader—what those engineers and operations guys did was incredible."

With a few exceptions, Marcus accomplished this without opposition from environmentalists or others. She met with her Heal the Bay buddies from time to time, but in her mind the adversarial relationships begun by Howard Bennett, and sometimes continued by Rim Fay, were over. "We had won," she says now. "It was total conquest. . . . There wasn't anything to push back on, because we were all doing it together by then. I mean, I don't think there was anything that we disagreed on."

Finally, on schedule, Hyperion went online with full secondary treatment, on November 23, 1998.

About the same time, Rim Fay was out diving in the bay when he started feeling weak and a little dizzy. He surfaced, struggling to swim back to his boat, the *Torpedo*. He was having a stroke. Fortunately, a Harbor Patrol boat came by, and he was able to wave it over to pick him up.

The strokes continued. "Rim never really recovered from that," Don May says sadly. "He just kind of went downhill." Some people believe his health problems came from his diving for long hours, going too deep, and coming up too fast. "Rim probably got bends at times, I would bet," says Martin Byhower, who knew Fay in the early 1980s and was a diver himself. "I'll bet he overdid it."

Others thought it was his alcoholism. "I went to parties at his lab," Moe Stavnezer says. "He'd make a tub of seviche for everyone, and there was never, ever a shortage of alcohol. And I can't say that I didn't imbibe myself, because I did. But Rim had a drinking problem. I mean, at times it probably led to some of his outbursts and to some of his intransigence on certain things. But on the whole, even despite that, Rim was smarter than most people in the world."

In 2002, at the Long Beach Aquarium, Rim Fay entered an anniversary party as a hero of sorts. Thirty years before, in 1972, voters had approved Proposition 20 and created the California Coastal Commission. Fay helped write the legislation and was one of its first—and most irascible—commissioners, fighting just about anything that involved development along the coast. That lack of compromise, his friends say now, got him fired. But on this night, the same confrontational attitude was celebrated as the kind of integrity few people could maintain.

After he walked into the room with a little help from friends Janet Bridgers and Alan Sanders, the cute girl at the registration table practically squealed, "Oh, you're Rim Fay! You're a legend!" It didn't matter that the seventy-three-year-old Fay smelled slightly of fish or seemed a little off after several strokes. The crowd welcomed him as an environmental champion.

"He was always self-effacing about that kind of thing," Bridgers says now, "because Rim was so centered on what it takes to make something happen. He always came to hearings wearing overalls and stuff like that. He just referred to himself as a fisherman. Not a scientist, not a PhD. He never flaunted his credentials."

Becoming more and more incapacitated, Fay ended up in Berkeley West Convalescent Hospital in Santa Monica. Patrick Wall visited him until Fay no longer recognized him. Wall's mentor was reduced to lying in bed, curled up in a fetal position. "You know," Wall says, "he was on painkillers, but he didn't want to be there. I think he wanted to get out and kill himself." Finally, on January 1, 2008, the seventy-eight-year-old Rim Fay died of a heart attack.

As his friend Alan Sanders puts it now, "People who do know about his work really treasure his memory so much. He really put himself out there. In essence, he sacrificed his ability to work in his field because he was so notorious working these public trust issues. Those kinds of sacrifices are rare."

CHAPTER 17

The 50 Percent Job

When a scoop of dark, sandy mud comes up from the seafloor some 180 feet below the research boat *La Mer*, it looks like a gelatinous blob of lifeless dirt. Looking at it, one can't help but wonder if the infamous "dead zone" so debated in 1985 still exists. Nevertheless, two marine biologists for the City of Los Angeles' Environmental Monitoring Division eyeball the dripping pile as though they're about to open a Christmas present. "New mud!" one laughs. "Never disturbed."

The two, with a combined fifty-three years' experience, dump the shiny glob into a large metal sink with almost childlike delight, as if they can't wait to see what critters might emerge from the mud. With water pumped from the sea, they gently spray down the mud, and slowly life separates out. At first, there's just "shell hash," the broken-up detritus of clams, snail shells, and scaphopods known as tusk shells. But then worms, the color of a tongue, show up, slowly writhing as though in shock. One scientist identifies them as *Cerebratulus californiensis*, a

(Top) Los Angeles marine biologists start to extract a sample scooped from the Santa Monica Bay seafloor in 2008.

(Bottom) Dirt sample from the seafloor before it is washed down and marine life specimens are extracted.

(Top) Seafloor sample after the dirt has been washed away. The stringlike squiggles are brittle stars, an indication that life is coming back to polluted areas.

(Bottom) Bringing up another sample at the stern of *La Mer.*

fairly common worm in the bay's sediments. "They probably get to full length, probably two feet when fully extended," he says.

Red filaments start to appear from the sand and pebbles, looking so fragile they can't possibly be attached to anything alive. As more mud disappears through the screen it was placed on, these filaments turn out to be several brittle stars, a spindly type of starfish whose presence often indicates clean—or at least, *cleaner*—water. Years ago the brittle stars moved out when Hyperion spewed mostly wastewater that had undergone only primary treatment. The fact that they're coming back, albeit slowly, shows full secondary treatment has helped create a more livable environment for these sensitive creatures.

One of the scientists points to a gray, marblelike animal barely discernible in the mud and says it's a sea grape. "It's a bioturbate. That's a good sign," he adds. Bioturbation occurs when animals such as the sea grape spend their lives rooting about the sediment, creating vertical tubes and caverns. In the case of Santa Monica Bay, where, in places, the organic material from the sewage treatment plants has settled into an immovable mat, industrious creatures such as the sea grape move about the sediment just enough that little bits of it float off and are suspended in the water. That's actually a good thing, because in this example dilution *is* the solution. Eventually, given enough sea grapes or other similar creatures poking about the seafloor, the remaining material that has discouraged species diversity could in theory disappear.

The specimens are taken to a bench on the starboard side. While kelp flies swarm around his head, a scientist bottles the critters in a propylene phenoxitol solution that "relaxes" them before he later kills them in formalin, a solution of formaldehyde. He separates out the burrowing anemones, which react with the relaxant, leaving a goo of sorts on the other animals.

Once the research crew docks later that afternoon, they'll take the animal specimens—along with water and sand samples—to Hyperion's lab, where two staff taxonomists will identify the species and record the information. Other scientists will then compare the findings to those of previous years for a picture of how the environment has changed, for better or worse. The water and sand samples are analyzed for chemicals, metals, and other toxicants.

This monitoring is mandated as part of the City of Los Angeles' permit to discharge treated wastewater into the bay. Indeed, the EPA and the Regional Water Quality Control Board regulate every aspect of the specimen collection. And while some of the samples today will also form part of a once-every-five-years survey that extends along the entire Southern California coast, the biologists routinely visit forty-four different Santa Monica Bay sampling sites or stations throughout the year. Using a crane, they drop their ninety-two-pound grabber and dig out sediment samples, as well as collect water several yards below the surface. Twice a year they trawl the bay for fish and large invertebrate animals.

And based on a few lively buckets of mud, the marine environment some two miles offshore is looking pretty good. That's not to say the bay is as pristine as Rim Fay once claimed it was when he was a kid. According to Curtis Cash, the research vessel's supervisor on this particular morning, one of the latest concerns is PBDEs (polybrominated diphenylethers), a class of flame-retardant chemicals that is somehow getting into the environment. While no one knows exactly how it's happening, according to the EPA it could occur during the manufacture of textiles and plastics by a process that uses the chemical to stifle combustion. Certainly, the PBDEs that make it to the sewer system easily slip past the secondary treatment and go out into the bay.

The water and sediment samples they're collecting this day are also being analyzed for pyrethroids, pesticides used for, among other things, killing mosquitoes that might carry the sometimes fatal West Nile virus.

As Cash is talking, the crane brings up another grab of mud. One of the scientists opens a hatch to the container and jabs a ruler inside to measure the depth to ensure the scoop meets field operations requirements for an adequate sample. Later, while they go through the sand and gravel, one mutters, "Tons of animals."

It's almost as if the bay has become its old fecund self, fully recovered from the 1985 pollution levels. Ask Heal the Bay, however, and they'll say the job of healing the bay is only half-way complete, though their Web site doesn't explain how they came up with that notion. They do detail any number of grim issues remaining, but seem most concerned about storm water runoff washing the city's streets clean and collecting in storm drains that take the resulting urban slurry of oil, pesticides, metals such as lead and zinc, and other toxins directly into the Santa Monica Bay surf.

Like the DDT issue, storm water runoff hasn't quite lit up the public in the way the 301(h) waiver did. And the issue has been around a lot longer than just about anything else involving Santa Monica Bay pollution. The Los Angeles storm drain system, designed and built in the early 1900s, was purposely routed to the ocean rather than to Hyperion, which was never designed to handle the massive volume a heavy rainstorm can bring.

At the March 25, 1985, meeting of the Los Angeles Regional Water Quality Control Board that included the first tepid hearing on the 301(h) waiver, board members wrung their hands over storm water runoff. With a kind of almost impolitic language one wouldn't expect from a government agency, they declared that the problem seemed bigger to them than the primary-treated wastewater and sludge going into the bay. Betty Werthman, who wasn't shy about supporting the 301(h) waiver, said in reference to a proposed study to see what pollutants were going into the bay, "I think we might find that we might be surprised that there are some serious things going down the storm drain that we really don't have much control of."

In essence, Congress had the same thoughts, and timidly amended the Clean Water Act in 1987 to include urban runoff, but did it with vague requirements that called on cities to clean up the problem to the maximum extent possible using "best management practices," whatever that means.

"It's just basically: Do a whole bunch of different things," Mark Gold, now Heal the Bay's president, says. "Monitor the receiving water, and hopefully [it meets] standards. But you don't have this direct nexus between numeric effluent limit and water quality standards. You just have a requirement for water quality standards to be met. And so the end result has been [that] the progress has been negligible. You can't even demonstrate that the water coming from Ballona Creek or LA River or any of those sources is any cleaner today than it was back in 1990."

That's not to say *nothing* has been done. The city has installed low-flow diversion structures that direct urban runoff to Hyperion in dry weather. However, when the volume passes a certain point during a storm, gates automatically reroute the water to the ocean. Also, in 1991 Heal the Bay pushed the City of Santa Monica into passing an ordinance to treat or filter storm water runoff. Zev Yaroslavsky, now a Los Angeles County supervisor, lists a number of plans under way to clean up storm water runoff. "We're going for a countywide storm water program, which will be very ambitious, very comprehensive, and very expensive," he says. "But it has to be done, because [runoff] is what is contributing now to the pollution in the Santa Monica Bay." One completed project captures runoff from creeks coming out of the San Gabriel Mountains east of the San Fernando Valley (north of the city of Los Angeles) in a series of cisterns, where the water is filtered. From there, the water goes to spreading grounds, where it seeps into the ground and recharges the aquifer that supplies well water to the city. "This is going to be a template for how we deal with flooding problems and storm water runoff and filtration issues throughout the county. It's a lot less expensive than building a flood control channel. And it's environmentally much more sound," Yaroslavsky says.

It's safe to say that such reclamation projects are so low key that the public doesn't know they exist, but Heal the Bay—with its aim to make the issue as visible as possible—sent out volunteer Gutter Patrols to the city's sixty thousand storm-drain catch basins in 1994 to stencil warnings not to dump garbage into the drains. And even though these graphics still exist on city sidewalks, Gold concedes, "we have a long way to go."

Mas Dojiri, who manages Los Angeles' environmental monitoring division, isn't sure the bay has reached the halfway point to healing or not. It could be better; it could be worse. From a scientist's point of view, the answer is complicated by so many factors that it can't be quantified. Just the same, he isn't shy about saying what's good about, and what's wrong with, water quality in Santa Monica Bay; and the sometimes bacteria-laden runoff pouring from the drains into the ocean is one of the things that's currently wrong. "We really need to focus on storm water, not just for bacteria but for metals and organics and [other pollutants]," he says.

To that end, his staff weekly tests the surf along fifty miles of Santa Monica Bay shoreline for three indicator bacteria—total coliforms, enterococcus, and fecal coliforms *(E. coli)*—scooping out samples in ankle-deep water at thirty-eight beach stations. Given the concern over storm water, the collection points were shifted to within splashing distance of the twenty-five drains that deposit runoff at Santa Monica Bay beaches. This increased the chances of bacterial spikes in their results, and Heal the Bay has subsequently given the nearby beaches low to failing grades on occasion after stormy weather.

Dojiri is quick to note that half the data used for Heal the Bay's report cards comes from his staff's efforts. "You have to give some sort of credit to the division that's been monitoring there essentially every single day," he says. Without prompting, he adds, "After forty years of testing, we never found any evidence that indicated that the Hyperion discharge was coming onshore and causing bacterial exceedances. In fact, what we did find was, whenever there was bacterial exceedances of any one of the three indicator bacteria along the shoreline, it was always either very close to a flowing storm drain or during a rain event, where all this water would come out of the storm drain, and it flowed along the coast, and you would get hits at the various beaches."

With that in mind, Heal the Bay recommends waiting seventy-two hours before swimming in the ocean after a rainstorm, and with good reason. Epidemiological studies, including one that Gold cowrote with Robert Haile in 1985, have shown that swimming near an outlet exposes one to increased risks of respiratory disease and gastrointestinal illness. The study concluded that people who swam in surf near a storm drain were 50 percent more likely to contract colds, sore throats, gastroenteritis, and other illnesses than someone frolicking further away.

While the storm water issue is yet to be solved, the environment affected by Hyperion's wastewater discharge is far better off now that the city's sewage goes through full secondary treatment. And although the bay is not completely "healed," Dojiri says, "[full secondary treatment] actually increased the number of species there and the abundance there. It actually improved it better than I had predicted."

Using the delicate brittle star as an indicator animal because of its strong aversion to the organic-rich pollutants that come from primary-treated sewage, biologists are discovering that the tiny creatures are showing up in small numbers again around the five-mile outfall, once classified as "degraded" by SCCWRP scientists.

In other words, the Hyperion sewage treatment plant is doing its job. This isn't to say the bay has totally recovered from the days of sludge. The area around the seven-mile outfall where the plant discharged its sludge continues to repel most marine life sensitive to the high-organic, low-oxygen environment that the yard-thick sludge created.

"When [the environmental monitoring division] did a study on that, probably ten or so years ago, we predicted that, within five years from that study, it'll become more natural," Dojiri said in 2008. "It's a slower recovery than we had predicted." And there remains the matter of what he calls the "legacy pollutants"—that is, the DDT, PCBs, and other organic compounds coating a 6.6-square-mile area in the bay. The toxicants are still showing up in concentrations high enough to have an effect on wildlife. Anglers are still advised not to eat the bottom-feeding white croaker and to limit their meals of scorpionfish, rockfish, and kelp bass to once every two weeks. In other words, while simply turning off the spigot that previously gushed sludge and primary-treated sewage has had a major effect on the bay's marine environment, no one really knows how much more must be done.

Not that there aren't scientists out there digging into the problem. In a large, one-story building with an El Pollo Loco restaurant incongruently attached to its side, SCCWRP is still in business but without the public relations issues that nearly destroyed the organization in 1985. Operating out of cozy, carpeted offices and roomy labs at a Costa Mesa industrial park, the forty-six full-time staff now work for a seemingly volatile mix of dischargers and government regulators. The combination, however, has introduced enough self-restraint that the kind of pressure Willard Bascom once felt from dischargers to bring home the 301(h) waiver never befalls Dr. Stephen

Weisberg, SCCWRP's executive director since 1996. "It's [Bascom's] tainting of the organization that has led to the present structure," he says. "People had lost faith in what the organization did. And so what they did is invite the regulatory agencies in."

That more balanced structure involves a fourteen-member board, six of whose members belong to state and federal regulatory agencies, including the EPA and three different Regional Water Quality Control Boards (Los Angeles, San Diego, and Santa Ana). In terms of voting power, this gives them a little more clout than the four dischargers on the board, something Weisberg says the dischargers requested "because they wanted to make sure the place is viewed as being unbiased."

Also, money no longer has the influence it once did. When Weisberg took the job, he discovered that the dischargers still paid most of SCCWRP's $1.6 million budget. "I went to the regulators," he notes, "and said, 'Hey, guys. You're not paying your fair share.' And you know what? They said, 'Good point.'" In 2008, SCCWRP still received some of its $8.8 million budget from dischargers, but the bulk came from state and federal sources.

It's clear that Weisberg, a friendly, open man who comes to work wearing a floral-print shirt, jeans, and sneakers, would rather talk about SCCWRP's current projects than its resurrection. He takes a tour guide approach, occasionally dipping into scientific vernacular, but mostly sticks to a narrative that suggests he's talking to the local Kiwanis Club. "What we're all about is developing the tools that will be used five years from now to do [water quality] monitoring better," he says. With that practical goal in mind, SCCWRP has been tackling one of the biggest problems dischargers and regulators face—alerting the public to high bacteria counts at Santa Monica Bay beaches. The only feasible approach has been to collect water samples and see if the *E. coli*, total coliforms, and enterococcus bacteria that public health officials are most concerned about multiply in culture dishes. Unfortunately, scientists can't start counting the results for about twenty-four hours. As Weisberg puts it, "We tell people the next day you shouldn't have been swimming in the water two days ago. You're likely going to get sick. It's not a great warning system."

SCCWRP is studying a huge improvement over the culture method called quantitative polymerase chain reaction, or PCR, which forgoes growing the bacteria and merely measures the water samples for specific genes that indicate a bacteria presence. Not only is this faster—it takes one to two hours—but also the process is so accurate that it reveals exactly what kind of bacteria are swimming in the water. That is, in the past, high bacteria concentrations were sometimes found in water but later determined to be from, say, birds. And as unappealing as it sounds to be swimming in water spiced with avian feces, people don't necessarily get sick from the exposure. It's bacteria coming from humans that harms other humans. However, cultures grown in the lab don't always offer enough evidence to let biologists make the distinction. PCR does.

SCCWRP has also been investigating a somewhat hidden source of water pollution called atmospheric deposition. Pollution from cars and airplanes eventually settles out of the air and ends up in storm water runoff flowing to the bay. Researchers found that metals such as chromium, copper, lead, nickel, and zinc can make their way from air pollution to water pollution, and in fact, air pollution is responsible for about half of these contaminants that hit the surf as a result of runoff.

Another big issue is so-called emerging contaminants, or chemicals that aren't currently being measured but that could significantly harm the environment. These include pharmaceuticals, hair care products, industrial chemicals, and pesticides. SCCWRP is studying how to measure these contaminants and determining at what levels regulators should declare them a problem. "Of course, they all want the answer to that— regulators and dischargers equally," Weisberg says. "Then they know if they need to worry about it." Unfortunately, he adds, secondary treatment doesn't pull these chemicals out of the water; it's an engineering problem and not something SCCWRP scientists will address.

The bottom line for Weisberg is that this research is used for public policy, but unlike in the days when Bascom tried to influence how decisions were made, Weisberg merely presents the scientific conclusions and lets the regulators figure out the rest. If the bay is indeed half healed, it's up to agencies such as the EPA or Santa Monica Bay Restoration Commission to decide how to fix the other half.

Ask Tom Hayden if the bay is half clean or half polluted, and he might say the latter. Hayden, who's no longer in state government, complains about Chevron oil tankers still docking at the company's El Segundo refinery, not far from the Hyperion plant and within sight of anyone on the beach. In 1986, the Los Angeles Regional Water Quality Control Board filed suit against the company for dumping excessive waste into the bay, including ammonia, grease, and oil over five years, totaling 880 alleged violations. And in 1991, more than twenty-seven thousand gallons of light oil leaked from a tanker, the last such major spill at the facility.

"It's just a long, long story," Hayden says. "It's about corporate conflict management. Two issues: One is that Chevron wants those tankers, hundreds of them a year, in that bay. That's number one. Number two—the dischargers want a cost-effective way to keep using the bay to discharge their effluent. Now, how on earth could some sleazy lobbyist for an oil company and waste dischargers prevail politically in liberal, Democratic LA County, LA city, even the city of Santa Monica? That to me is the window into the limits of politics. Usually, we talk about the potential—that here a grassroots citizen movement [Heal the Bay] put this issue on the map and everything. That's the potential, the great potential of politics. But politics extracts its pound of flesh. There's movements and there's Machiavellians. The movements are transitory, and the Machiavellians are permanent.

"If [Rim Fay] were alive," he adds, "he would say we're not even close to what needs to be done. And it's a big problem, because the alleged success of the Santa Monica Bay cleanup is Xeroxed, duplicated, and sent forth to every bay and estuary in the United States and around the world, as if this is a model."

Indeed, rightly or wrongly, Heal the Bay is held up as one of the country's huge environmentalist successes. Perhaps *success* isn't the right word, given the halfway-healed argument. The group's *impact* has been great, infiltrating government and the public with what it means to pollute an important body of water and what it takes to clean it up. In that sense as well, the group is halfway there—the citizenry has been educated, but the bay still isn't clean. As Gold puts it, "You know, I feel like our progress on storm water has been embarrassingly poor." But as he told the *Los Angeles Times* in 1997, "From the standpoint of marine life in the bay, conditions are markedly improved." In other words, halfway there.

On May 19, 1989, Heal the Bay presented Howard Bennett with a plaque featuring its now customary fishbone logo and the simple inscription "In gratitude to Howard Bennett, who started it all." This could be interpreted in two ways. Bennett sees it as acknowledgment that his original coalition was the precursor to Heal the Bay, now one of the most visible—and perhaps most prominent—environmental groups in Southern California, if not the country. Indeed, on the occasional weekends when Heal the Bay organizes beach cleanups, he watches people picking up litter from the beach in front of his Playa del Rey home and claims this as one of his legacies.

On the other hand, the inscription has a cagey, elusive quality to it, as if avoiding actually naming Bennett as the group's founder and instead identifying him merely as the bottom rung, the catalyst to a movement that just happened to spawn Heal the Bay. "The role that Howard fulfilled is alerting people to the problem to begin with," Dorothy Green said. "That's important. But he was not at all influential, bottom line."

The early 2009 version of the Heal the Bay Web site didn't mention him at all. The group's beginning was described this way: "It was 1985, and a handful of people learned that the City of Los Angeles was dumping barely treated sewage into Santa Monica Bay. They learned that the pollution from sewage and storm drains had led to a decrease in the number and quality of fish in the Bay, dolphins that had reproductive problems and were

Howard Bennett in 2008. The schoolteacher brought the EPA 301(h) waiver to an entire city's attention.

full of tumors, a large patch of the bottom of the Bay was essentially lifeless, and people who swam and surfed in the Bay complained about infections and other illnesses. This handful of people, led by founder Dorothy Green, got angry and they got organized. Heal the Bay was born." (Later in 2009, when it was pointed out to Mark Gold, now Heal the Bay president, that Bennett had more to do with the group's beginnings than the Web site let on, the reference to Green was revised to read "founding president," but it still didn't credit the schoolteacher's contributions.)

There's a bit of truth and a bit of mythology mixed together in that grand statement. Green acknowledged that initially the "handful of people" she led were primarily the board members of the Los Angeles League of Conservation Voters, who first learned of the issue from, ahem, Bennett.

However, it isn't just Heal the Bay that's promulgated this image of themselves as the original do-gooders. In 1989, the *Los Angeles Times* referred to them as "the burgeoning volunteer organization that has successfully pressured the City of Los Angeles to stop dumping sewage sludge into the ocean." This was hardly the case, given that the EPA had been pestering the city for years over the sludge issue and, despite the fact that Los Angeles had missed deadline after deadline, the legal impetus for forcing Hyperion to stop discharging sludge was well on its way. Heal the Bay was more or less a witness to the process and certainly didn't have the political oomph at the time to pressure anybody.

But because Bennett largely disappeared from the fight after the Regional Water Quality Control Board nixed the 301(h) waiver, Green believed she was justified in erasing him from the official story. As she told Santa Monica's *Daily Breeze* in 1998,

"If we were going to win, somebody had to be around and see how it was going to be implemented and enforced. There was no staying power with Howard. He can't be responsible to other people. He has to do it himself."

Bennett doesn't deny the ephemeral nature of his involvement or his somewhat combative attitude, but maintains there's a bit more to it than a lack of long-term commitment. "When I get on something," he says now, "I don't know day from night. I'll get up at two o'clock in the morning and write notes. Forgetting it's no life for me, it's no life for [my wife,] Bente. It's no life for the family. And I was probably driving everybody crazy who knew me. Myself, too." As we've seen, he passed off his campaign to Green and went on vacation to shake off the stress. Bente Bennett adds that her husband was also obligated to devote his daytime hours to teaching, not demonstrating, and that's one more reason why he quit the campaign once they returned at the end of the summer.

And so the soon-to-be-formed Heal the Bay continued without him. As cynical as Howard Bennett can be at times, he prefers to tell his story with far more optimism. For him the story is a triumph—his triumph. As he sees it, this is a tale of how some schoolteacher took on the city, state, and federal governments and got them to clean up the sewage. And, as he quickly adds, his story has an educational and inspirational component: "'If that old man can fight and make a positive change that will help the world,'" he says, "'I can do it, too!'" And yet, he tacitly agrees with Hayden, or is it Heal the Bay? Either way, the job is only half done. The bay is still half polluted:

On July 26, 2008, 1,464 gallons of raw sewage spilled from a West Los Angeles pipeline clogged by fats, oil, and grease, known in the sewer biz as FOG. The Bureau of Sanitation's Wastewater Collection Systems Division collected all but 366 gallons and returned it to the system. The remaining sewage worked its way down Ballona Creek and into the ocean not far from Bennett's home, where his son, Leif, and Leif's son and daughter played in the surf.

The next morning, the Los Angeles Department of Public Health posted warning signs before Dojiri's monitoring crew could assess whether the bacteria levels were high enough to pose a public health hazard. According to Dojiri, they found the bacteria had been diluted so much they didn't need to worry, and the signs quickly disappeared.

About the same time, Leif came down with a crushing ear infection that required a strong antibiotic to cure. It's hard to say if some overachieving bug from the spill managed to find Leif's ear canal, or if it came from another source. In either case, Bennett alternated between bouts of fear for his son's health and anger that—twenty-three years after what he likes to call the "nine-month miracle"—one still can't jump in the ocean without any risk of disease.

Just the same, a few days later, the seventy-eight-year-old Bennett donned his black Speedo and swam just beyond the surf line, still in love with the ocean, still afraid of its power.

Epilogue

I last saw Dorothy Green when we met at her home on June 2, 2008, for a few follow-up interview questions and for her to pose for a portrait for this book. Even though she felt exhausted after a trip to San Francisco to promote a book she had written about water use issues, she gamely answered my questions (still showing a little irritation over any discussion of Howard Bennett) and even worked the camera like a pro when we took the pictures. A few days later, I sent her three pictures she had requested, and she wrote back, saying, "Thanks, Bill. And congratulations on getting the book off. Can't wait to see it."

Unfortunately, she never did. That September, Felicia Marcus told me Green was "failing fast." A melanoma that had first been diagnosed in the 1970s had metastasized in her brain in about 2003. And now the cancer had finally left her in hospice care, about to die. Despite the great effort it took simply to concentrate on any one thought for long, she dictated an article about water use and worked on a fund-raising strategy for the California Water Impact Network, a group she helped found.

According to Jamie Simons, at about the same time, several people from the original Heal the Bay group had gathered around her bed to tell stories while Green lay there in a fog (she had been out of it for days) when she suddenly awoke, completely clearheaded, and started adding to the stories, correcting people on their versions. But finally, a few days later, Dorothy Green died on October 13 at age seventy-nine.

Resources

PROLOGUE. SURFER SCIENTIST

Dorsey, John. Interview by author. February 12, 2008.

Dorsey, John, et al. *Santa Monica Bay Monitoring Study.* Hyperion Treatment Plant annual report, 1984.

U.S. Environmental Protection Agency. "National Estuary Program." Environmental Protection Agency Web site, www.epa.gov/nep/programs/smb.htm, accessed May 2008.

CHAPTER 1. THE SWIMMER

Bennett, Howard. Interviews by author. November 11, 2006, and December 19, 2007.

Citron, Alan. "Report Confirms Toxic Dumping; Hayden Decries Damage to Bay." *Los Angeles Times,* March 28, 1985.

Ferrell, David. "The 10-Year Battle of Santa Monica Bay." *Los Angeles Times,* May 13, 1985.

"Weather Report." *Los Angeles Times,* March 28, 1985.

CHAPTER 2. THE WITNESS

"Beach Areas Still under Quarantine." *Los Angeles Times,* June 3, 1951.

Byhower, Martin. Interview by author. November 21, 2007.

California Water Quality Control Board, Los Angeles Region. Meeting agenda. March 25, 1985.

———. Meeting audiotapes. March 25, 1985.

———. Meeting minutes. March 25, 1985.

City of Los Angeles. "Hyperion Sewage Treatment Plant." City of Los Angeles Department of Public Works, Bureau of Sanitation Web site, www.lacity.org/san/wpd/siteorg/general/hypern1.htm, accessed June 17, 2008.

Clean Water Act (Federal Water Pollution Control Act). Public Law 95-217. Amended November 27, 2002. EPA.

Dojiri, Mas. Division Manager, City of Los Angeles Department of Public Works, Bureau of Sanitation. Email correspondence with author. June 17, 2008.

Dorsey, John. Email correspondence with author. March 28, 2008.

Eklund, Patricia. Interview by author. September 16, 2008.

Fay, Rimmon C. "Dirty Water: The Personal Memoir of a Marine Biologist." *LA Weekly,* June 7, 1985.

———. Résumé. January 1987.

Ferrell, David. "Waiver's Reversal Shows Bureaucratic Infighting." *Los Angeles Times,* January 6, 1986.

Ghirelli, Robert. Interview by author. May 21, 2008.

"LA Urged to Abolish Public Works Board." *Los Angeles Times,* June 20, 1952.

Los Angeles City Administrative Office. *Economic and Demographic Information.* Officer report. City of Los Angeles, October 2, 2008.

May, Don. Email correspondence with author. February 27, 2008.

———. Interview by author. November 29, 2007.

Miele, Robert. Interview by author. March 4, 2009.

Moore, Joe. *Overview of Federal Water Quality Laws and Regulations.* New Mexico Water Resources Research Institute, November 1990.

"More Sewage Grief Seen by Engineer." *Los Angeles Times*, August 7, 1951.

Morrison, Patt. "Sea of Sludge Off Southland Charged." *Los Angeles Times*, January 19, 1974.

O'Reilly, Richard. "Sewage—How Much of It Can the Oceans Absorb?" *Los Angeles Times*, September 8, 1982.

"Q&A." Interview with Rimmon Fay. *Los Angeles Herald Examiner*, March 15, 1985.

Sklar, Anna. *Brown Acres: An Intimate History of the Los Angeles Sewers.* Santa Monica: Angel City Press, 2008.

Stewart, Jill. "Santa Monica Bay Blues." *Los Angeles Times*, August 17, 1986.

Tarvyd, Ed. Interview by author. January 10, 2008.

Toufexis, Anastasia. "The Dirty Seas." *Time*, August 1, 1988.

Turhollow, Chuck. Email correspondence with author. March 5, 2009.

———. "Total Maximum Daily Loads." U.S. Environmental Protection Agency Web site, www.epa.gov/owow/tmdl/intro.html, accessed January 15, 2008.

U.S. Environmental Protection Agency. "Amendments to Regulations Issued, the Clean Water Act, Section 301(h) Program." Environmental Protection Agency Web site, www.epa.gov/owow/oceans/discharges/301h.html, accessed June 17, 2008.

Wall, Patrick. Interview by author. January 7, 2008.

Weiss, Kenneth R. "Rimmon C. Fay, 1929–2008, Diver, Marine Scientist Fought Santa Monica Bay Pollution." *Los Angeles Times*, January 4, 2008.

"White Croaker." Pier Fishing in California Web site, www.pierfishing.com, accessed March 23, 2008.

CHAPTER 3. THE COALITION

Bennett, Bente. Interview by author. February 29, 2008.

Bennett, Howard. Interviews by author. October, 23, 2006, October 28, 2006, November 4, 2006, November 11, 2006, and December 19, 2007.

——. Notes by author taken during unrecorded conversations. N.d.

Bridgers, Janet. Interview by author. February 5, 2008.

Lansford, Ruth. Interview by author. December 19, 2007.

Rempel, William C., and Dale Fetherling. "Coastal Unit Halts Threat to Homes." *Los Angeles Times*, November 6, 1975.

CHAPTER 4. SQUIRP

Bascom, Willard. *The Crest of the Wave*. New York: Harper & Row, 1988.

——. "The Purpose of SCCWRP." Meeting notes. October 7, 1984.

Brown, David. Interviews by author. April 24, 2008, April 25, 2008, April 30, 2008, and June 9, 2008.

California Department of Fish and Game. Letter to Willard Bascom. March 7, 1979.

Fay, Rimmon C. "Dirty Water: The Personal Memoir of a Marine Biologist." *LA Weekly*, June 7, 1985.

O'Reilly, Richard. "Sewage Not Harming Ocean, Study Finds." *Los Angeles Times*, November 21, 1982.

Pace, Eric. "Willard Bascom, 83, Scientist and Leader in Deep-Sea Exploration." *New York Times*, October 10, 2000.

Powell, M. A., and G. N. Somero. "Sulfide Oxidation Occurs in the Animal Tissue of the Gutless Clam, *Solemya Reidi*." *Biological Bulletin* (August 1985).

Robak, Warren. "Quality of Coastal Waters Improving." *Torrance Daily Breeze*, December 5, 1982.

Sattoria, T. L. "Dumping Sludge in Ocean Not Harmful, Speakers Say." *Long Beach Press-Telegram*, July 29, 1983.

Sklar, Anna. *Brown Acres: An Intimate History of the Los Angeles Sewers*. Santa Monica: Angel City Press, 2008.

Steinman, David. "Sick Bay." *Los Angeles Easy Reader*, August 1, 1985.

Thompson, Bruce. Interview by author. June 5, 2008.

"Willard Bascom Dies; Explored Oceans for Science, Treasures." *Washington Post*, October 16, 2000.

CHAPTER 5. THE PRESS CONFERENCE

"Alaska Hitch-Hike Takes Only 17 Days." *New York Times*, August 29, 1951.

Arrendell, Stephen. "Coalition's Gripe: Hear Us Out on Bay." *Santa Monica Evening Outlook*, April 4, 1985.

Bennett, Howard. "Coalition to Stop Dumping More Raw Sewage into the Ocean." Press conference text. April 4, 1985.

———. Interviews by author. October 23, 2006, October 28, 2006, November 4, 2006, November 11, 2006, and December 19, 2007.

Bennett, Leif. Interview by author. November 19, 2007.

"Canada's Border Toughest to Cross." *Vancouver Sun*, May 7, 1952.

Citron, Alan. "Warning Issued on Fish Caught Off Southland." *Los Angeles Times*, April 12, 1985.

Cohen, Paul. "Student Wins Bet, Hitches to Alaska." *The Ticker*, September 17, 1951.

Goodson-Todman Productions. Letter to Howard Bennett. September 4, 1951.

Haefele, Marc. Interview by author. January 29, 2009.

"Hemisphere Jaunt Logs 30,000 Miles." *New York Times*, June 22, 1952.

KABC-TV. Report on the Coalition to Stop Dumping Sewage into the Ocean press conference. *Eyewitness News*, evening report, April 4, 1985.

KCBS-TV. Report on the Coalition to Stop Dumping Sewage into the Ocean press conference. *Channel Two News Live at 5*, April 4, 1985.

May, Don. Email correspondence with author. May 27, 2008.

"Student Hitch-Hikes from New York to Anchorage." *Anchorage Daily News*, August 28, 1951.

U.S. Department of State. Letter to Howard Bennett. August 18, 1952.

"Who Was David Brower?" David Brower Center Web site, www .browercenter.org, accessed February 29, 2009.

CHAPTER 6. CITY HALL

Bennett, Leif. Interview by author. November 19, 2007.

Coalition to Stop Dumping Sewage into the Ocean. "A History of Irresponsibility, Los Angeles, and Sewage Dumping in Santa Monica Bay." Press release. N.d.

"The Dilemma of the Bay." Editorial. *Los Angeles Herald*, August 18, 1985.

Dojiri, Mas. Division Manager, City of Los Angeles Department of Public Works, Bureau of Sanitation. Email correspondence with author. February 24, 2009.

Ferrell, David. "L.A. Pressured to Treat Bay Sewage Fully." *Los Angeles Times*, May 16, 1985.

———. "The 10-Year Battle of Santa Monica Bay." *Los Angeles Times*, May 13, 1985.

"Fouling Our Own Nest." Editorial. *Los Angeles Herald*, May 16, 1985.

Gage, Mike. Interview by author. June 6, 2008.

Galanter, Ruth. Interview by author. January 17, 2008.

Kindel, Maureen. Interview by author. February 11, 2008.

Marcus, Felicia. Interview by author. February 9, 2008.

Morrison, Patt. "Sea of Sludge Off Southland Charged." *Los Angeles Times*, January 19, 1974.

O'Reilly, Richard. "Sewage—How Much of It Can the Oceans Absorb?" *Los Angeles Times*, September 8, 1982.

Robak, Warren. "Plan to Ease Restrictions at Hyperion Plant Challenged." *Torrance Daily Breeze*, May 14, 1985.

Sklar, Anna. *Brown Acres: An Intimate History of the Los Angeles Sewers.* Santa Monica: Angel City Press, 2008.

Tetra Tech. *Technical Evaluation of Application for Modification of the Requirements of Secondary Treatment, Hyperion Treatment Plant, City of Los Angeles.* Los Angeles: Tetra Tech, February 1981.

Turhollow, Chuck. Email correspondence with author. March 5, 2009.

Yaroslavsky, Zev. "Stomping Mad over Sewage Disposal." *Los Angeles Times*, December 19, 1976.

CHAPTER 7. THE ACTIVIST

Bennett, Howard. Interview by author. December 19, 2007.

Citron, Alan. "Toxic Fish, County Won't Post Warnings." *Los Angeles Times*, April 19, 1985.

Green, Dorothy. Email correspondence with author. June 4, 2008.

———. Interviews by author. January 16, 2008, and June 2, 2008.

———. Letter from the Los Angeles League of Conservation Voters to the Environmental Quality Commission and Board of Public Works. April 11, 1985.

Gustaitis, Rasa. "Dorothy Green and the Power of Water." *Coast & Ocean* (Spring 2006).

Los Angeles League of Conservation Voters. "Mission Statement." Los Angeles League of Conservation Voters Web site, www.lalcv .org, accessed June 4, 2008.

Marcus, Felicia. Interview by author. February 9, 2008.

Parent, Randi. "Dorothy Green: A Life of Volunteerism." *Currents: The Newsletter of Heal the Bay* (Spring 2005).

Stavnezer, Moe. Interview by author. February 6, 2008.

U.S. Environmental Protection Agency. "Earth Day." Environmental Protection Agency Web site, www.epa.gov/earthday.history.htm, accessed June 3, 2008.

CHAPTER 8. THE SECOND HEARING

Bennett, Howard. Interviews by author. October 23, 2006, October 28, 2006, November 4, 2006, November 11, 2006, and December 19, 2007.

California Water Quality Control Board, Los Angeles Region. Meeting agenda. May 13, 1985.

———. Meeting audiotapes. May 13, 1985.

———. Meeting minutes. May 13, 1985.

Coalition to Stop Dumping Raw Sewage into the Ocean. "New Hearing Granted by EPA on Sewage Dumping." Press release. April 11, 1985.

Dorsey, John. Email correspondence with author. March 5, 2009.

———. Interview by author. February 12, 2008.

Eco News. Television report on Water Quality Control Board hearing. May 13, 1985.

Eklund, Patricia. Interview by author. September 16, 2008.

Ferrell, David. "The 10-Year Battle of Santa Monica Bay." *Los Angeles Times,* May 13, 1985.

Ghirelli, Robert. Interview by author. May 21, 2008.

Green, Dorothy. Interviews by author. January 16, 2008, and June 2, 2008.

Marcus, Felicia. Email correspondence with author. May 8, 2008.

———. Interview by author. February 9, 2008.

Rethlake, Kathy. "Hyperion Hearing Slated." *Santa Monica Evening Outlook,* April 11, 1985.

Robak, Warren. "Plan to Ease Restrictions at Hyperion Plant Challenged." *Torrance Daily Breeze,* May 14, 1985.

U.S. Environmental Protection Agency. Letter to Howard Bennett. April 9, 1985.

Wall, Patrick. Interview by author. January 7, 2008.

———. "Ocean Sewage Dumping Runs Afoul." *Los Angeles Times,* May 12, 1985.

CHAPTER 9. THE SCIENTIST

"Alcan's Effluent Suspect in Marine Life Changes." *Victoria Times Colonist,* September 12, 1980.

Associated Press. "Ocean Pollution Said Widespread." May 17, 1984.

Bascom, Willard. *The Concept of Assimilative Capacity: SCCWRP Annual Report.* Southern California Coastal Water Research Project, 1981–82.

———. "Material Presented to Assemblyman Tom Hayden's Group." Southern California Coastal Water Research Project. May 17, 1985.

———. "The Purpose of SCCWRP." Memo by Willard Bascom. Southern California Coastal Water Research Project, October 7, 1984.

———. *SCCWRP Biennial Report*. Southern California Coastal Water Research Project, 1983–84.

———. "Was the Emperor Crazy?" *Sea Technology* (June 1984).

Brown, Anne. Interview by author. April 30, 1985.

Brown, David. "Contamination of Coastal Southern California." Written presentation for the State Assembly Task Force Investigation of Toxic Pollution in Santa Monica Bay hearings. May 17, 1985.

———. Interviews by author. April 24, 2008, April 25, 2008, April 30, 2008, and June 9, 2008.

———. Letter to Tom Hayden, May 18, 1985.

"Chemical Wastes Off Peninsula Still Pose Threat." *San Pedro News Pilot*, February 28, 1984.

Citron, Alan. "Report Confirms Toxic Dumping: Hayden Decries Damage to Bay." *Los Angeles Times*, March 28, 1985.

Crust, John. "Off-Catalina Toxic Dump Called Worst in the World." *Los Angeles Herald Examiner*, February 28, 1985.

Garlington, Phil. "Ex-Water Quality Board Staffers Say Bosses Impeded Inspections." *Los Angeles Herald Examiner*, May 18, 1985.

———. "White Croaker Earns Name as Fish Is Found Full of DDT." *Los Angeles Herald Examiner*, February 27, 1985.

Hayden, Tom. "Assembly Task Force to Hold Hearings on SM Bay Pollution." Press release. May 14, 1985.

———. Letters to David Brown. May 6, 1985, and May 15, 1985.

Jones, Jack. "Ocean Site Reportedly Used for Toxics Dump." *Los Angeles Times*, February 28, 1985.

KABC-TV. Report on toxics dump. *Eyewitness News*, evening report. February 29, 1985.

KCBS-TV. Report on toxics dump. *Channel Two News Live at 5*. February 27, 1985.

———. Report on water board hearings. *Channel Two News Live at 5*. March 1, 1985.

Manisco, Patricia. "Marine Dump: A Garden or an Eco-Mess?" *Los Angeles Times*, July 15, 1984.

Miele, Robert. Interview by author. March 4, 2009.

Morgenthaler, Anne. "DDT Research Ordered Hushed?" *Santa Monica Daily Breeze*, May 18, 1985.

O'Reilly, Richard. "Pollution along Coast Surprises Scientists." *Los Angeles Times*, May 16, 1984.

———. "Sewage—How Much of It Can the Oceans Absorb?" *Los Angeles Times*, September 8, 1982.

Rethlake, Kathy, and Donna Prokop. "Santa Monica Bay Tainted with Toxics." *Evening Outlook*, February 28, 1985.

———. "S.M. Bay Pollution Called among the Worst." *Torrance Daily Breeze*, March 1, 1985.

Rossi, Mitchell S. "DDT: The Plague That's Stalking California's Ocean." *San Diego Newsline*, January 16, 1985.

Sklar, Anna. *Brown Acres: An Intimate History of the Los Angeles Sewers*. Santa Monica: Angel City Press, 2008.

Smith, Doug. "Coastal Waters Improving, Study Shows." *Los Angeles Times*, February 1, 1981.

Southern California Coastal Water Research Project. *Southern California Coastal Water Research Project Directors and Staff by Discipline, 1969–2000*. Los Angeles: Southern California Coastal Water Research Project, n.d.

Stammer, Larry B., and Lee Dye. "Most Pollution Off Coast Laid to Sewage, Not Ocean Dumps." *Los Angeles Times*, March 1, 1985.

State Assembly Task Force for the Investigation of Toxic Pollution in Santa Monica Bay. Agenda. May 17, 1985.

Thompson, Bruce. Interview by author. June 5, 2008.

U.S. Congress. House of Representatives. Committee on Public Works and Transportation, Subcommittee on Water Resources. *Modification of Secondary Treatment Requirements for Discharges into Marine Waters*. Los Angeles, May 24, 25, 1978.

U.S. Environmental Protection Agency. "Commencement Bay—Nearshore Tideflats." Environmental Protection Agency Web site, www.yosemite.epa.gov/R10/cleanup.nsf, accessed on June 8, 2008.

————. "Superfund." Environmental Protection Agency Web site, www.epa.gov/superfund/about.htm, accessed June 10, 2008.

CHAPTER 10. THE POLITICIAN

Bascom, Willard. "Santa Monica Bay on the Mend." *Los Angeles Times*, July 3, 1985.

Bennett, Howard. Interview by author. November 21, 2006.

Brill, Judy. "Raw Sewage in Ballona Creek Ends Up in Marina, PdR Waters." *Argonaut*, July 18, 1985.

Citron, Alan. "Are Bay Fish Safe to Eat? Showdown Expected Tuesday." *Los Angeles Times*, May 27, 1985.

————. "Report Confirms Toxic Dumping; Hayden Decries Damage to Bay." *Los Angeles Times*, March 28, 1985.

————. "Researcher Hid Severity of Bay Contamination, Aide Charges." *Los Angeles Times*, May 22, 1985.

————. "Water Research Director Denies Aide's Charges of Bay Pollution." *Los Angeles Times*, May 23, 1985.

"Controversial Head of Water Project on Leave Till Retirement." *Los Angeles Herald Examiner*, May 31, 1985.

Dorsey, John. Interview by author. February 12, 2008.

Green, Dorothy. Interview by author. January 16, 2008.

Hayden, Tom. Email correspondence with author. September 7, 2008.

————. Interviews by author. January 14, 2008, and February 4, 2008.

————. "Legislation." Tom Hayden Web site, www.tomhayden.com/legislation, accessed May 12, 2008.

————. *Rebel*. Los Angeles: Red Hen Press, 2003.

Keller, Larry. "Panel Backs L.B. Firm on Fish Pollution." *Long Beach Press Telegram*, May 29, 1985.

Morgenthaler, Anne. "DDT Research Ordered Hushed?" *Santa Monica Daily Breeze*, May 18, 1985.

Rainey, James. "Sewage Spill in Creek Hit by Hayden." *Los Angeles Times*, July 14, 1985.

Rethlake, Kathy. "Chemists Debate Toxic Effects on Fish." *San Pedro News-Pilot*, May 29, 1985.

———. "Scientists Clear Water Research Chief." *Santa Monica Daily Breeze*, May 31, 1985.

———. "S.M. Bay Pollution Called among the Worst." *Santa Monica Daily Breeze*, March 1, 1985.

Schmidt, Bob. "Fish Pollution Issue Revives; L.B. Firm's Report Disputed." *Long Beach Press Telegram*, May 22, 1985.

Southern California Coastal Water Research Project. Scientists' press release. May 28, 1985.

Stambler, Lyndon. "Funds Allocated for Report on Santa Monica Bay Marine Life." *Los Angeles Times*, October 4, 1984.

Statement by SCCWRP Blue-Ribbon Panel distributed to press organizations. May 30, 1985.

Steinman, David. "Fish Stories: 'Data Excellent,' Panel Tells Public." *LA Weekly*, June 7–13, 1985.

Thompson, Bruce. Interview by author. June 5, 2008.

CHAPTER 11. THE BROWN RIBBON

Bascom, Willard. "Santa Monica Bay on the Mend." *Los Angeles Times*, July 3, 1985.

Bennett, Bente. Interview by author. February 29, 2008.

Bennett, Howard. Interview by author. November 4, 2006.

———. Letter to James Grossman. June 6, 1985.

Bennett, Leif. Interview by author. November 19, 2007.

Earth Alert! Patrick Wall biography. Earth Alert! Web site, www .earthalert.org/about_us.html, accessed January 6, 2008.

Ferrell, David. "EPA May Grant Extension for Dumping in Santa Monica Bay." *Los Angeles Times*, July 7, 1985.

Hayden, Tom. Letter to the editor. *Los Angeles Times*, July 13, 1985.

Pyrillis, Rita. "Hyperion Plant Protesters March on City Hall." *Santa Monica Daily Breeze*, June 3, 1985.

"The Region." Brief report on brown ribbon demonstration. *Los Angeles Times*, June 3, 1985.

Wall, Patrick. Interviews by author. November 4, 2006, and January 7, 2008.

CHAPTER 12. HEAL THE BAY

Bennett, Bente. Interview by author. February 29, 2008.

Bennett, Howard. Interview by author. December 19, 2007.

———. Letter to the editor. *Coast Lines*, a newsletter of the Sierra Club Angeles Chapter Clean Coastal Waters Task Force. July 1985.

Bridgers, Janet. Interview by author. February 5, 2008.

"Bring Back the Beach." *Currents: The Newsletter of Heal the Bay* (Summer 2005).

Byhower, Martin, Heal the Bay. Letter to California Regional Water Quality Control Board, Los Angeles Region, comments on NPDES permit. N.d.

Green, Dorothy. Interviews by author. January 16, 2008, and June 2, 2008.

Haefele, Marc. Interview by author. January 29, 2009.

Heal the Bay. "Milestones." Heal the Bay Web site, www.healthebay .org/aboutus/20years/milestones.asp, accessed November 15, 2007.

———. "Who We Are." Heal the Bay Web site, www.healthebay.org/ aboutus.asp, accessed November 15, 2007.

Los Angeles League of Conservation Voters. Memo to Coalition to Stop Dumping Sewage into the Ocean. October 28, 1985.

———. "The Public Is Angry and Organizing." Letter. September 1985.

Marcus, Felicia. Interview by author. February 9, 2008.

May, Don. Interview by author. November 29, 2007.

Simons, Jamie. Interview by author. February 6, 2009.

Stavnezer, Moe. Interview by author. February 6, 2008.

Wall, Patrick. Interview by author. January 7, 2008.

CHAPTER 13. THE DIRTY TOILET AWARDS

Antonovich, Michael. Letter to Howard Bennett. November 5, 1985.

Arrendell, Stephen. "Coalition Dumps on Bradley." *Santa Monica Evening Outlook*, November 6, 1985.

Bennett, Howard. Interviews by author. November 4, 2006, November 11, 2006, and December 19, 2007.

———. Text for "Dirty Toilet Awards." N.d.

Bennett, Leif. Interview by author. November 19, 2007.

Coalition to Stop Dumping Sewage into the Ocean. "Dirty Toilet Awards." Press release. N.d.

"'Dirty Toilet Award' to Mayor, City Council over Sewage Dumping." *Argonaut*, November 7, 1985.

Ferrell, David. "EPA May Grant Extension for Dumping in Santa Monica Bay." *Los Angeles Times*, July 7, 1985.

Green, Dorothy. Interviews by author. January 16, 2008, and June 2, 2008.

Haefele, Marc. Interview by author. January 29, 2009.

KCOP-TV. Report on Dirty Toilet Awards. *Evening News*. November 5, 1985.

Richard, Chris. "Sewage Treatment Draws New Attack." *Culver City News*, November 14, 1985.

Simons, Jamie. Interview by author. February 6, 2009.

CHAPTER 14. THE DECISION

Bennett, Howard. Interviews by author. November 4, 2006, November 11, 2006, and December 19, 2007.

Boyarsky, Bill. "Sewer Spills a Political Peril for Bradley?" *Los Angeles Times*, October 1, 1985.

"Bradley Changes Position on Wastewater Treatment." *Argonaut*, September 12, 1985.

Bradley, Tom. Letter to James Grossman, California Regional Water Quality Control Board. September 4, 1985.

California Water Quality Control Board, Los Angeles Region. Meeting audiotapes. November 25, 1985.

———. Meeting minutes. March 25, 1985.

Chandler, John. "Sewage Spills Cost City a Record $150,000 Fine." *Los Angeles Herald Examiner,* October 29, 1985.

Decker, Cathleen. "City Weighs Its Option to Modernize Sewer System." *Los Angeles Times,* September 4, 1985.

———. "L.A. Agrees to Pay $30,050 Fine for Raw Sewage Spills into Ocean." *Los Angeles Times,* August 22, 1985.

———. "Report Urges Stricter Rule on Treatment of Sewage." *Los Angeles Times,* November 19, 1985.

Decker, Cathleen, and Janet Clayton. "Council Decides to Fight Fine for Sewage Spillage." *Los Angeles Times,* October 24, 1985.

Fanucchi, Kenneth. "Water Board May Fine L.A. for Spills at Ballona Creek." *Los Angeles Times,* July 25, 1985.

Ferrell, David. "Waiver's Reversal Shows Bureaucratic Infighting." *Los Angeles Times,* January 6, 1986.

Ghirelli, Robert. Interview by author. May 21, 2008.

Green, Dorothy. Interview by author. January 16, 2008.

Hayden, Tom. Interview by author. January 14, 2008.

Kindel, Maureen. Interview by author. February 11, 2008.

Marcus, Felicia. Interview by author. February 9, 2008.

Morgenthaler, Anne. "New Ballona Sewage Spill Reported." *Santa Monica Evening Outlook,* July 24, 1985.

———. "Raw Sewage Again Flows into Ballona Creek." *Santa Monica Evening Outlook,* July 23, 1985.

———. "Sewage Spills Fine." *Santa Monica Evening Outlook,* August 9, 1985.

Rainey, James. "Sewage Spill in Creek Hit by Hayden." *Los Angeles Times,* July 14, 1985.

Regional Water Quality Control Board. Meeting minutes. November 22, 1985.

Sklar, Anna. *Brown Acres: An Intimate History of the Los Angeles Sewers.* Santa Monica: Angel City Press, 2008.

State Water Resources Control Board. Letter to Tom Bradley. October 28, 1985.

Steinman, David. "Water Board Postpones Action on Hyperion Treatment Waiver." *Los Angeles Easy Reader*, July 11, 1985.

Stewart, Jill. "A Gamble on Sewage Treatment That Went Sour." *Los Angeles Times*, January 6, 1986.

U.S. Environmental Protection Agency. Letter to Tom Bradley. September 30, 1985.

———. "U.S. EPA Announces Final Denial of Wastewater Treatment Waiver." Press release. March 12, 1986.

CHAPTER 15. FRIEND OF THE COURT

Bennett, Howard. Interview by author. November 4, 2006.

City of Los Angeles Board of Public Works, Bureau of Sanitation. *Initial Evaluation Report, Hyperion Treatment Plant*. City of Los Angeles, February 1986.

Coalition to Stop Dumping Sewage into the Ocean. "Has Santa Monica Bay Taught Mayor Bradley a Lesson?" Press release. N.d.

Deukmejian Campaign Committee. "Largest Polluter." Radio ad. October 1986.

Estrada, Ray. "Local Teacher Says Deal Dupes Public." *Culver City Wave*, October 8, 1986.

Flore, Faye. "Court Allows Groups to Fight Sea Dumping." *Torrance Daily Breeze*, May 30, 1986.

Gage, Mike. Interview by author. June 6, 2008.

Ghirelli, Robert. Interview by author. May 21, 2008.

Green, Dorothy. Interviews by author. January 16, 2008, and June 2, 2008.

Kindel, Maureen. Interview by author. February 11, 2008.

Koenenn, Connie. "An Environmentalist for All Seasons Series." *Los Angeles Times*, June 16, 1986.

Marcus, Felicia. Email correspondence with author. May 8, 2008.

———. Interview by author. February 9, 2008.

Marcus, Felicia, and Joel Reynolds. Letter to James Grossman, California Regional Water Quality Control Board, written on behalf of the Center for Law in the Public Interest. June 7, 1985.

Merina, Victor. "City to Fully Treat Ocean Sewage." *Los Angeles Times*, December 18, 1985.

Prokop, Donna M. "Water Board Reassigns Top 2 Hyperion Plant Officials." *Santa Monica Daily Breeze*, December 19, 1985.

Roderick, Kevin. "City, EPA Reach Accord on Sewage-Dumping Suit." *Los Angeles Times*, July 31, 1986.

Simons, Jamie. Interview by author. February 6, 2009.

Stammer, Larry B. "EPA to Study Santa Monica Bay Waste for Superfund Aid." *Los Angeles Times*, December 17, 1985.

Stewart, Jill. "Pollution Solution Activists Vow to Bird-Dog Every Step of Santa Monica Bay Cleanup Series." *Los Angeles Times*, April 12, 1986.

Sullivan, Patricia. "Anne Gorsuch Burford, 62, Dies; Reagan EPA Director." *Washington Post*, July 22, 2004.

Taylor, Robert E. "Los Angeles Sets Sewer Spending in EPA Accord." *Los Angeles Times*, October 1986.

CHAPTER 16. OUTSIDERS AND INSIDERS

Bay, Steven. Interview by author. February 23, 2009.

Bennett, Howard. "Has Santa Monica Bay Taught Mayor Bradley a Lesson?" Coalition to Stop Dumping Sewage into the Ocean press release. N.d.

———. Interviews by author. November 4, 2006, November 11, 2006, and December 19, 2007.

Bridgers, Janet. Interview by author. February 5, 2008.

Brown, David. Interviews by author. April 24, 2008, April 25, 2008, April 30, 2008, and June 9, 2008.

Byhower, Martin. Interview by author. November 21, 2007.

Deukmejian Campaign Committee. "Largest Polluter." Radio ad, October 1986.

Gage, Mike. Interview by author. June 6, 2008.

Galanter, Ruth. Interview by author. December 19, 2007.

Gold, Mark. Interviews by author. January 8, 2008, and February 4, 2008.

Green, Dorothy. Interviews by author. January 16, 2008, and June 2, 2008.

Hayden, Tom. Interview by author. January 14, 2008.

Marcus, Felicia. Interview by author. February 9, 2008.

May, Don. Interview by author. November 29, 2007.

Merina, Victor, and Marylouise Oates. "Kindel Quits Key City Post; 4th to Leave in 3 Months." *Los Angeles Times*, August 27, 1987.

Miele, Robert. Interview by author. March 4, 2009.

Outwater, Alice B. *Reuse of Sludge and Minor Wastewater Residuals.* Boca Raton, FL: CRC, 1994.

Roderick, Kevin. "City Officials Plead Sewage Case on the Enemy's Turf." *Los Angeles Times*, April 7, 1986.

Sanders, Alan. Interview by author. January 26, 2009.

Santa Monica Bay Restoration Commission Web site, www .santamonicabay.org, accessed March 9, 2009.

Simons, Jamie. Interview by author. February 6, 2009.

Stavnezer, Moe. Interview by author. February 6, 2008.

Taylor, Nancy. Interviews by author. November 9, 2007, and November 20, 2007.

Thompson, Bruce. Interview by author. June 5, 2008.

Wall, Patrick. Interview by author. January 7, 2008.

CHAPTER 17. THE 50 PERCENT JOB

Abrahamson, Alan. "Bay Watch: More Fish, Cleaner Water." *Los Angeles Times*, August 11, 1997.

Author observation aboard *La Mer*, July 8, 2008.

Bennett, Howard. Interview by author. December 19, 2007.

Cash, Curtis. Email correspondence with author. August 4, 2008.

———. Email correspondence with author. September 30, 2008.

————. Interview by author. July 8, 2008.

Colford, John M., et al. *Water Quality Indicators and the Risk of Illness in Non-Point Source Impacted Recreational Waters*. Southern California Coastal Water Research Project, 2006.

Dojiri, Mas. Division Manager, City of Los Angeles Department of Public Works, Bureau of Sanitation. Email correspondence with author. August 1, 2008.

————. Interview by author. June 13, 2008.

Fiore, Faye. "Chevron Faces $8.8 Million Fine." *San Pedro News-Pilot*, August 22, 1986.

Gold, Mark. Interview by author. January 8, 2008.

Green, Dorothy. Interview by author. January 16, 2008.

Hayden, Tom. Interview by author. January 14, 2008.

Heal the Bay. "Accomplishments." Heal the Bay Web site, www .healthebay.org, accessed November 15, 2007.

————. "Swimming in the Bay." Heal the Bay Web site, www.healthebay .org/stayhealthy/swimming/default.asp, accessed November 15, 2007.

————. "Who We Are." Heal the Bay Web site, www.healthebay.org/ aboutus.asp, accessed November 15, 2007.

Koenenn, Connie. "An Environmentalist for All Seasons Series: Reshaping the Future." *Los Angeles Times*, June 16, 1989.

Los Angeles Regional Water Quality Control Board. Meeting audio-tapes. March 25, 1985.

Rabin, Jeffrey L. "Making a Splash: Officials Beginning to Listen to Heal the Bay's Campaign for Cleanup Effort." *Los Angeles Times*, February 8, 1990.

Radcliffe, Jim. "The Sky's the Limit." *Santa Monica Daily Breeze*, June 22, 1998.

Sabin, Lisa D., and Kenneth C. Schiff. *Metal Dry Deposition Rates along a Coastal Transect in Southern California*. Southern California Coastal Water Research Project, 2007.

Weisberg, Stephen. Interview by author. February 23, 2009.

Yaroslavsky, Zev. Interview by author. February 9, 2009.

EPILOGUE

Green, Dorothy. Email correspondence with author. June 27, 2008.

Lopez, Steve. "Cancer Can't Dim Passion for a Cause." *Los Angeles Times*, September 17, 2009.

Marcus, Felicia. Email correspondence with author. September 18, 2009.

Simons, Jamie. Interview by author. February 6, 2009.

Woo, Elaine. "Environmentalist Began Heal the Bay." *Los Angeles Times*, October 14, 2009.

Index

Text: 10/15 Janson
Display: Janson
Compositor: Westchester Book Group
Printer and Binder: Maple-Vail Book Manufacturing Group